Ultrasonic Techniques for Fluids Characterization

ULTRASONIC TECHNIQUES FOR FLUIDS CHARACTERIZATION

MALCOLM J. W. POVEY
Department of Food Science
University of Leeds
Leeds, United Kingdom

ACADEMIC PRESS
San Diego London Boston New York Sydney Tokyo Toronto

This book is printed on acid-free paper. ∞

Copyright © 1997 by ACADEMIC PRESS

All Rights Reserved.
No part of this publication may be reproduced or transmitted in any form or by any means, electronic or mechanical, including photocopy, recording, or any information storage and retrieval system, without permission in writing from the publisher.

Permissions may be sought directly from Elsevier's Science and Technology Rights Department in Oxford, UK. Phone: (44) 1865 843830, Fax: (44) 1865 853333, e-mail: permissions@elsevier.co.uk. You may also complete your request on-line via the Elsevier homepage: http://www.elsevier.com by selecting "Customer Support" and then "Obtaining Permissions".

Academic Press
An Imprint of Elsevier
525 B Street, Suite 1900, San Diego, California 92101-4495, USA
http://www.apnet.com

Academic Press Limited
24-28 Oval Road, London NW1 7DX, UK
http://www.hbuk.co.uk/ap/

Library of Congress Cataloging-in-Publication Data

Povey, M. J. W. (Malcolm J. W.)
　　Ultrasonic techniques for fluids characterization / by Malcolm J.W. Povey.
　　　　p.　　cm.
　　Includes bibliographical references and index.
　　ISBN-13: 978-0-12-563730-5
　　1. Ultrasonic waves—Industrial applications.　2. Sonochemistry.
I. Title.
TP156.A33P68　1997
541.3'45'0287—dc21　　　　　　　　　　　　　　97-9788
　　ISBN-13: 978-0-12-563730-5　　　　　　　　　　　CIP

Transferred to Digital Printing 2009

Contents

Preface xi
Acknowledgments xiii

1

Introduction

1.1 The Beginnings 1
1.2 Understanding Sound 2
1.3 Representations of Sound 3
1.4 Sounds Classical and Sounds Quantum 5
1.5 Comparisons between Light and Ultrasound 6
1.6 The Adiabatic Idealization 7
1.7 Common Sense Is Unsound 8
1.8 Scope of This Work 9
1.9 How to Use This Book 9

2
WATER

2.1 Measurement of Sound Velocity 11
   2.1.1 Introduction 11
   2.1.2 Accuracy and Errors 14
      2.1.2.1 Temperature 15
      2.1.2.2 Acoustical Delays 15
      2.1.2.3 Impedance 16
      2.1.2.4 The Control of Reverberation with Buffer Rods 18
      2.1.2.5 Acoustical Bonds 19
      2.1.2.6 Power Levels 19
      2.1.2.7 Diffraction and Phase Cancellation 20
      2.1.2.8 Timing Errors Due to Trigger Point Variation 23
      2.1.2.9 Measuring Group Velocity 24
   2.1.3 Calibration 25
2.2 The Dependence of Velocity of Sound on Density and Compressibility 25
   2.2.1 The Velocity of Sound in Mixtures and Suspensions 26
   2.2.2 The Velocity of Sound in Air/Water Mixtures 28
   2.2.3 The Importance of Removing Air from Samples 29
   2.2.4 The Effects of Temperature on Propagation in Water 29
   2.2.5 The Effects of Pressure on Propagation in Water 31
   2.2.6 Sound Velocity in Equidensity Dispersions 31
2.3 The Relationship between Velocity and Attenuation—Conditions of High Attenuation 33
2.4 The Compressibility of Solute Molecules 34
   2.4.1 Introduction 34
      2.4.1.1 Empirical and Semiempirical Methods 34
      2.4.1.2 Concentrations 35
   2.4.2 Determining Partial Volumes 36
      2.4.2.1 The Method of Intercepts 36
   2.4.3 Apparent Molar Quantities 37
      2.4.3.1 Apparent Specific Volume 37
      2.4.3.2 Apparent Compressibility 38
      2.4.3.3 Concentration Increments 39
   2.4.4 The Dilute Limit 39
      2.4.4.1 Partial Specific Volume and Partial Specific Adiabatic Compressibility 40
   2.4.5 Sound Velocity and Concentration—The Urick Equation 40
   2.4.6 Determining the Compressibility of Solute Molecules—a Summary 42
   2.4.7 Experimental Data on Compressibility and Its Interpretation 44
      2.4.7.1 Protein 44

3
MULTIPHASE MEDIA

3.1 Apparatus 47
3.2 Determining Composition in the Absence of Phase Changes 49
    3.2.1 Alcohol 50
    3.2.2 Sugar 51
    3.2.3 Concentration of a Dispersed Phase in a Colloidal Phase 52
    3.2.4 Analysis of Edible Oils and Fats 52
    3.2.5 Cell Suspensions 54
    3.2.6 Temperature Scanning 54
3.3 Following Phase Transitions 54
    3.3.1 General Comments 54
    3.3.2 Attenuation Changes 56
    3.3.3 Crystallizing Solids 57
    3.3.4 Crystallization in Colloidal Systems 59
3.4 Determination of Solid Fat Content 60
    3.4.1 Introduction 60
    3.4.2 General Method 61
        3.4.2.1 Region I 61
        3.4.2.2 Region III 62
        3.4.2.3 Region II 63
    3.4.3 Margarine 65
    3.4.4 Chocolate 65
    3.4.5 Accuracy 68
    3.4.6 Anomalies Close to the Melting Point 69
    3.4.7 Comparison with Dilatometry and Pulsed Nuclear Magnetic Resonance 70
    3.4.8 Solid Content and Particle Size 72
3.5 Crystal Nucleation 72
    3.5.1 Crystal Nucleation Rates 72
    3.5.2 Ice 74
3.6 The Solution–Emulsion Transition and Emulsion Inversion 75
    3.6.1 Emulsion Inversion 76
3.7 Determination of Emulsion Stability by Ultrasound Profiling 76
    3.7.1 Introduction 76
    3.7.2 History 77
    3.7.3 The Leeds Profiler 78
    3.7.4 Interpretation of Ultrasound Velocity Profiles 80
        3.7.4.1 Renormalization 81
        3.7.4.2 Limits of Applicability of Renormalization Method 83
    3.7.5 Examples of Profiling 84
3.8 Summary 90

4

Scattering of Sound

4.1 Theories of Sound 91
4.2 A Comparison of Electromagnetic and Acoustic Propagation 92
4.3 Scattering Theory 93
    4.3.1 Why Scattering Theory? 93
    4.3.2 What Is Scattering? Assumptions of Scattering Theory 94
        4.3.2.1 Long Wavelength Limit 99
        4.3.2.2 Low Attenuation 100
        4.3.2.3 Plane Wave 100
        4.3.2.4 Scattering Is Weak 100
        4.3.2.5 Random Distribution of Particles 100
        4.3.2.6 Adiabatic Approximation 100
        4.3.2.7 Navier–Stokes' Form for the Momentum Equation 101
        4.3.2.8 Thermal Stresses Neglected 102
        4.3.2.9 No Changes in Phase 102
        4.3.2.10 Linearization of Equations 102
        4.3.2.11 Temperature Variations 102
        4.3.2.12 System Is Static 102
        4.3.2.13 Particles Are Spherical 102
        4.3.2.14 Infinite Time Irradiation 103
        4.3.2.15 Pointlike Particles 103
        4.3.2.16 No Overlap of Thermal and Shear Waves 103
        4.3.2.17 No Interactions between Particles 103
        4.3.2.18 Lack of Self-Consistency 103
    4.3.3 A Description of Weak Scattering 104
        4.3.3.1 Wave Potentials 104
        4.3.3.2 Modes in a Pure Liquid 106
        4.3.3.3 Thermoelastic Scattering 109
        4.3.3.4 Viscoinertial Scattering 110
        4.3.3.5 Scattered Waves Combine within the Transducer 110
    4.3.4 Plane Wave Incident on a Single Particle 111
        4.3.4.1 Introduction 111
        4.3.4.2 Spherical Harmonics 111
        4.3.4.3 Boundary Conditions 112
    4.3.5 Scattering by Many Particles 116
        4.3.5.1 Introduction 116
        4.3.5.2 Multiple Scattering Theories 116
    4.3.6 Numerical Calculations Using Scattering Theory 118
        4.3.6.1 Particle Size Distribution and Change in Phase 118
    4.3.7 The Results of Scattering Theory 118
    4.3.8 Simplified Scattering Coefficients 120
    4.3.9 Working Equations 124

        4.3.9.1 The Urick Equation   124
        4.3.9.2 The Multiple Scattering Result   124
        4.3.9.3 The Modified Urick Equation   125
        4.3.9.4 Experimental Determination of the Scattering Coefficients   127
   4.3.10 Multiple Dispersed Phases   129
   4.3.11 MathCad Calculation Results   130
   4.3.12 Experimental Validation of Acoustic Scattering Theory   133
4.4 Scattering from Bubbles   137

5

ADVANCED TECHNIQUES

5.1 Particle Sizing   141
   5.1.1   Introduction   141
   5.1.2   Review   142
   5.1.3   Theoretical Limitations of Acoustic Particle Sizing   143
   5.1.4   Relaxation Effects   144
   5.1.5   Ultrasonic Methods of Particle Sizing   145
        5.1.5.1 Simultaneous Measurement of Velocity and Attenuation   145
        5.1.5.2 Determination of Particle Size from Velocity and Attenuation   145
        5.1.5.3 Bandwidth and Signal-to-Noise Ratio   147
        5.1.5.4 A Particle Sizing Apparatus—Pulsed Method   148
        5.1.5.5 Continuous-Wave Interferometer   148
        5.1.5.6 Commercial Particle Sizing Apparatus   149
        5.1.5.7 Electroacoustics   150
        5.1.5.8 The Future—Measurement Systems   152
5.2 Propagation in Viscoelastic Materials   152
   5.2.1   Introduction   152
   5.2.2   Measuring Aggregation in Viscoelastic Materials   156
        5.2.2.1 Introduction   156
        5.2.2.2 Detecting Aggregation with Ultrasound Profiling   157
        5.2.2.3 Computer Modeling   158
        5.2.2.4 Aggregation of Casein   159
   5.2.3   Frequency-Dependent Ultrasound Profiling   161
   5.2.4   Particle Size Effects in Ultrasound Profiling   162
5.3 Bubbles and Foams   163
5.4 Automation and Computer Tools   163
   5.4.1   The Computer as Controller   165
   5.4.2   Windows   165
   5.4.3   Prototyping   167
   5.4.4   RS232C   168
   5.4.5   IEEE Bus   168
   5.4.6   Instrument Programming   169

5.4.7 *Oscilloscope 169*
       5.4.7.1 Fourier Analysis 169
5.4.8 *Timer–Counter 172*
5.4.9 *The UVM 173*
5.4.10 *Transducer Excitation 173*
5.4.11 *Cabling 173*
5.4.12 *Calibration 174*
5.4.13 *Sample Changer 174*
5.4.14 *Temperature Control 174*
5.4.15 *Data Storage and Analysis 174*
5.4.16 *Conclusion 175*

APPENDIX A
Basic Theory 177
A.1 The Isotropic Elastic Solid 178
A.2 The Pure Newtonian Fluid 179
A.3 Mixture of Two Pure Newtonian Fluids 179
A.4 Viscoplastic Materials 179

APPENDIX B
MathCad Solutions of the Explicit Scattering Expressions 181

GLOSSARY 185
BIBLIOGRAPHY 191
INDEX 205

PREFACE

Ultrasonic velocimetry has reached a stage of development today that is comparable to that of nuclear magnetic resonance a decade ago. Ultrasound can penetrate optically opaque materials, providing high-quality information at low cost on bulk properties in a rapid and noninvasive manner. It readily lends itself to on-line measurement. The technique has something to offer all scientists and engineers and deserves serious examination by anyone concerned with automated analysis and measurement.

Ultrasound velocity measurement has many virtues for process monitoring. It can be accurate, fast, reliable, and relatively low-cost. In some cases, ultrasound velocity is the only feasible way of determining process parameters.

The considerable advances made in this area over the past decade are brought together, for the first time, in one volume. Developments in physical acoustic and ultrasonic methods are increasingly taking place outside the traditional physical and electronic engineering environments within which they were first developed. Disparate disciplines now employ these techniques and have developed them further. Because each discipline has developed ultrasound techniques as a tool to another end, not as an end in itself, these developments often have not been communicated beyond a small circle.

This volume describes recent developments in the understanding of ultrasonic propagation through liquid and viscoelastic materials and the application of these advances to the characterization of dispersions, emulsions, solutions, and soft solids. It provides a comprehensive and up-to-date account of the field and covers topics such as ultrasonic profiling, crystallization of lipids, particle sizing, and rapid composition determination.

Measurement of the velocity of sound is the only direct way to determine adiabatic compressibility. This measurement opens the door to a new world of materials characterization, a world that hitherto has remained the province of specialists. For the biochemist, this book shows how to determine protein hydration; for the food scientist, to follow changes in solid fat content; for the physical chemist, to measure solute–solute and solute–solvent interactions; for the medic, to measure blood cell aggregation.

The book gives a clear and novel exposition of the basics of ultrasonic propagation. New applications of the technique in the reader's own field of interest will become clear. Both the newcomer and advanced practitioners in ultrasound will benefit from this book. A separate treatment of advanced theory points the reader to the appropriate computer tools. Full details of experimental techniques are given. This handbook contains everything that the reader needs to begin sound velocity measurement.

This book is intended for scientists and engineers working in academia, in the food, chemical, biochemical, pharmaceutical, cosmetic, petroleum, and oil process industries, medicine, and in the life sciences. It may be used as the basis for a final-year undergraduate or first-year postgraduate course in ultrasonics.

Malcolm J. W. Povey

Acknowledgments

A work such as this stands on the shoulders of years of outstanding scientific effort, especially that made since the 17th century. Yet, science has only scraped the surface of the possibilities open to human society.

The author owes a debt to the very many students and collaborators with whom he has worked on the subject of ultrasound over the years. Without their contributions, which have been acknowledged at relevant points in this work, this book could not have been written. Blandine Buttner, Christian Seiler, Karlo Hering, and Yongtao Wang provided valuable feedback using drafts of the chapter on multiphase media as a laboratory manual.

Special thanks go to Dr. Valerie Pinfield and Dr. Julian McClements, who acted as my reviewers. Valerie Pinfield's detailed comments enormously improved this work. Her thesis work also formed the foundation for the chapter on scattering theory. Julian McClements' discovery of the importance of thermal scattering in emulsion systems, while he was working as a research student with me, opened my eyes to the possibilities inherent in this phenomenon. Of course, all errors are my responsibility.

Finally, it is conventional to acknowledge the contribution of one's family. But this does not lessen their contribution. Without the support of my partner, Pat Jones, I could not have written this work.

1

INTRODUCTION

1.1 THE BEGINNINGS

It is very easy to measure the velocity of sound. All you need is a stopwatch and a measuring tape. In 1827 Colladon and Sturm used a bell to produce a ring of sound in the waters of Lake Geneva. They used their ears as receiver and a stopwatch to time the clap of the bell after it traveled through the lake. They determined the distance between two boats, one on each side of the lake, with a tape measure and used the flash from a shot of gunpowder to signal the initial clap of the bell. In this way they found the velocity of sound in the waters of the lake. You can dispense with the colleague and the gunpowder if you can find a suitable canyon. Place yourself with your back to the canyon wall, then shout and start the stopwatch simultaneously. Stop the stopwatch when you hear your echo. Then measure the width of the canyon, double the distance, and divide it by the time on the stopwatch to obtain the velocity of sound.

The idea that sound is a wave phenomenon grew out of observations of water waves (Pierce, 1981a; Lindsay, 1973, in which many of the original papers referred to in subsequent paragraphs are reprinted). This idea is mentioned by the Greek philosopher Chrysippus (ca. 240 B.C.), by the Roman architect and engineer Vetruvius (ca. 25 B.C.), and by the Roman philosopher Boethius (A.D. 480–524). The wave idea was also promulgated by Aristotle (384–322 B.C.). However, the ancients did not have the technical apparatus necessary to exploit their ideas.

Newton (1642–1727) in his *Principia* (1686) first expounded the mechanical interpretation of sound as being "pressure" pulses transmitted through neighboring fluid particles. The foundations for the present classical theory of sound propagation were laid by Euler (1707–1783), Lagrange (1736–1813), and d'Alembert (1717–1738), who gave continuum physics a definite mathematical structure. These developments coincided with the increasing availability of technical apparatus to permit their exploitation—especially the clock, gun, and telescope (Hessen, 1931).

In the last century, sound velocity measurement was the standard method of determining the adiabatic compressibility of materials and hence the ratio of specific heats (Zemansky, 1957a); it was therefore a little-noticed but indispensable part of the science of thermodynamics. It has remained so to this day.

Today, sound velocity measurement is ubiquitous in industrial processes; go around any process plant and you are certain to find ultrasonic devices for level, concentration (see Chapter 2), and flow measurement (Sheppard, 1994) in liquids and gases. In metal fabrication, ultrasound nondestructive evaluation (NDE) and testing (NDT) are important quality control tools (Kuttruff, 1991b). Ultrasonic NDE is characterized today by frequencies between 0.1 and 50 MHz, pulsed operation and low power levels (<1 W m^{-2}). In civil engineering, ultrasound velocity measurement has its part to play in monitoring the porosity of concrete and the integrity of metal structures. The safety of ships depends on ultrasonic determination of the corrosion layer of ships' hulls.

Medical applications of ultrasound usually do not involve measurement of sound velocity, but instead depend on the relative invariance of sound velocity in human tissue. Many medical apparatuses measure the reflected signal and display the spatial variation of its amplitude, often using the time domain to give depth to the image (Kuttruff, 1991c). Blood flow is measured using ultrasound doppler (Fox, 1996).

The majority of the current industrial applications of ultrasound depend on the development from radar principles of the pulse-echo technique by Pellam and Galt (1946). Pellam and Galt's work forms the technical basis for most of the equipment described in this work.

The existing applications of ultrasound scrape the surface of what we can learn from this form of motion. To get a measure of the possibilities it is only necessary to look to nature. The bat actually "sees" with ultrasound (Suga, 1990). With ultrasound echo ranging and backscatter, different species of bat can locate their quarry, identify and select the type of insect, and then catch it in the presence of obstacles, all at high velocity. With the metaphor of "color" (relating to frequency variation) in mind, imagine what would be possible if we could "see" with ultrasound. Recent advances in technology make this objective a perfectly feasible one, and aspects of the technique are widely used in medical scanning and visualization.

1.2 UNDERSTANDING SOUND

Much of the theoretical basis for the analysis of sound velocity in liquids used here is found in Wood's *A Textbook of Sound*, which was republished in 1964 (Wood, 1941a, 1964). Wood provides experimental and technical flesh for the bare bones of the theoretical work of Lord Rayleigh (Strutt, 1872, 1877, 1896) in the 19th century and Lamb in the early 20th century (Lamb, 1925), among many others. The theory of sound is firmly based in classical physics, in particular Newton's laws of motion, and most especially his second law, which relates force to acceleration through *force = mass × acceleration*. Newton developed a theory relating the velocity of sound to the compressibility and density. Laplace (1816) pointed out that the compressibility in this relationship was adiabatic, not isothermal as Newton thought. The theory of sound, as developed by Rayleigh, transcends the second law by its wavelike nature. It is interesting to note that Strutt (1872) discovered a famous result during his investigation of the scattering of sound. He showed that the sky is blue because of the fourth-power dependence on wavelength for scattering from small particles.

1.3 REPRESENTATIONS OF SOUND

As has already been stated, attempts to understand sound have drawn on analogies with water waves. Pictures and analogies are a very powerful method for understanding sound. Wave representations of sound give a picture of the collective motion of large numbers of particles, whose individual motions still need to be recognized (Pierce, 1981a). This work is mainly concerned with a special case of sound propagation in which the particle motions which constitute the pressure pulses are parallel to the direction of propagation of the pulse. This is called *compression* or *longitudinal* sound. Unless stated otherwise, we shall always mean this type of sound wave. A representation of compression sound is given in Figure 1.1.

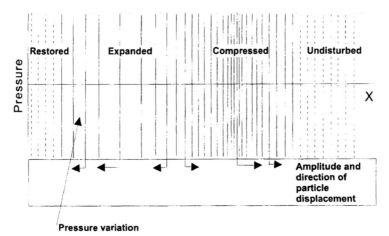

FIGURE 1.1 Particle position, particle displacement, and spatial pressure variation plotted against position (x) for a single cycle, sinusoidal, plane-traveling wave in a fluid medium. (Adapted from Pierce, 1981a.)

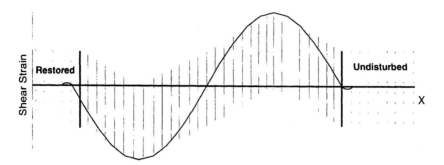

FIGURE 1.2 Particle position, particle displacement, and spatial shear strain variation plotted against position (x) for a single cycle, sinusoidal, plane-traveling shear wave.

In contrast, the shear mode of propagation of sound is supported over macroscopic distances in solids but not in liquids. In the shear mode, particle displacement is transverse to the direction of propagation (Figure 1.2). There are other possible modes of acoustic propagation, but these are of little relevance to this work.

In nearly every application of ultrasound to be discussed here, a special type of ultrasound source, called the piston source, is used. Where wave phenomena are concerned, the nature of the source and detector cannot be separated from the nature of the interaction of sound with the material under investigation. The performance of the transducer depends on its acoustical matching to the medium under investigation, as well as its electromagnetic matching to the detection and generation circuitry. The sound field in front of a piston source (Figure 1.3) may

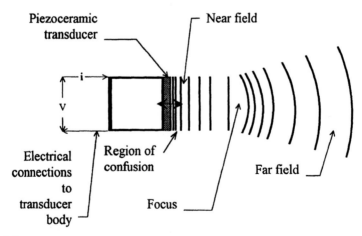

FIGURE 1.3 Piston source operating with the diameter (aperture) very much greater than the acoustic wavelength in the medium in contact with the front of the transducer.

be represented using an analogy with water waves first employed by Newton to describe sound. In this representation, lines are drawn which coincide with the crests and troughs of the wave (this method of representing sound waves is used in Figure 1.1 as well). These lines are also lines of constant *"phase."* The *phase* of a wave refers to the position of maximum or minimum in a wave (the peaks and troughs just referred to), relative to the waveform as a whole.

It can be seen from Figure 1.3 that the sound field in front of the piston source is a very complicated one. The implications of this will be considered in Chapter 2.

1.4 SOUNDS CLASSICAL AND SOUNDS QUANTUM

The theory in this book does not venture beyond the bounds of classical physics. However, developments in quantum theory, most especially in the quantum theory of heat have established that sound is but one aspect of molecular motion whose spectrum extends to at least 10^{13} Hz (Kittel, 1971) in solids. The wave–particle dichotomy applies to sound waves as it does to all forms of motion in matter. In the classical picture which has been adopted in this work, however, only the wave aspects of sound are considered.

Classical mechanics is the attempt to explain the motions of material bodies by means of Newton's Laws of Motion. This explanation is generally successful, provided bodies are not too small (larger than atomic dimensions of 0.1 nm) and not moving too fast (less than 1/10 of the speed of light, 3×10^8 m s^{-1}). Quantum mechanics were developed to account for the properties of small objects, and general relativity to account for motion close to the speed of light (Taylor, 1970).

All motions of a body, apart from translation, may be termed sound of one sort or another. These motions make up the total heat capacity of insulating materials and are a part of the heat capacity of noninsulators. The heat capacity in insulators always approaches zero as the temperature approaches zero, and this can only be explained if their vibrations are quantized, a discovery due to Einstein (Kittel, 1971). The quantization of sound is central to the modern theory of heat, which is one of the great triumphs of quantum theory. The discussion in this work is confined to the lowest frequencies ($<10^9$) and longest wavelengths (>1 μm) of the total sound frequency spectrum (Figure 1.4).

Ultrasound is high frequency with regard to human hearing (>18 kHz) but it may be transformed into much higher frequency sound waves (phonons) whose behavior can only be understood as being both that of a wave and that of a particle. Phonons are the most common type of sound (Figure 1.4). Although it is beyond the scope of this book, a number of important material testing methods arise from the exploitation of very high-frequency sound, for example, Brillouin scattering (Beyer and Letcher, 1969) and heat pulse methods (Almond and Patel, 1996).

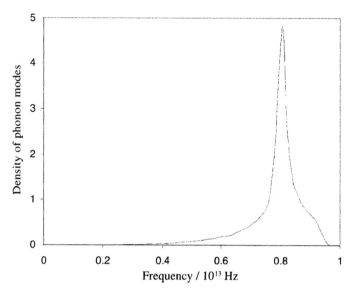

FIGURE 1.4 The density of phonon modes plotted against frequency for compression sound in aluminum. (Adapted from Kittel, 1971.)

1.5 COMPARISONS BETWEEN LIGHT AND ULTRASOUND

Despite having fundamentally different origins, light and sound have many similarities because of their common wave/particle duality. A mathematical apparatus has been developed over the years which has benefited each area of investigation in turn. Rayleigh's investigations of sound led to his realization that scattering of light was the cause of the blueness of the sky. Jones (1986) is a standard reference for anyone interested in the basic mathematical apparatus for the description of acoustic and electromagnetic waves. A comparison of some of the features of the two types of radiation appears in Table 1.1.

Light detectors, including our eyes, are generally phase insensitive. This means that it is possible for us to see on a cloudy day, when the light reaching the ground from the sun has been thoroughly jumbled by scattering by the water droplets which constitute clouds. On the other hand, ultrasound transducers are phase sensitive and cannot detect a signal in which the phases between different parts of the signal have been mixed up. As a result, bubbly liquids are generally opaque to ultrasound, whereas light is much less affected by the presence of small bubbles in water. The phase sensitivity of ultrasound transducers necessitates much greater care in the design of apparatus than the wavelengths involved imply. However, the phase sensitivity of ultrasound transducers make it easy to ensure coherence between successive pulses. *Coherence* means that the phase relationship between one pulse and another is preserved.

TABLE 1.1 Features of Ultrasound Which Distinguish It from Light Scattering in Water Based Colloids at 20°C

Ultrasound	Light
Transducers are phase sensitive	Transducers are phase insensitive
Wavelength between cm and μm	Wavelength between 0.5 and 1 μm
Frequency between 0.1 and 10^{13} Hz	Frequency between 3×10^{16} and 6×10^{16} Hz
Coherence between pulses	No coherence between pulses
Responds to elastic, thermophysical, and density properties	Responds to dielectric and permeability properties
Particle motion parallel to the direction of propagation; no polarization	Field displacement perpendicular to direction of propagation; polarization is therefore possible
Propagates through optically opaque materials	Sample dilution is normally required

Light radiation can be polarized whereas compression sound cannot. In solids it is possible to produce a mode of sound called *"shear"* (Figure 1.2) which can be polarized, but only fluids, in which the shear mode of propagation is highly attenuated, are considered in this work. It is easier to control both phase and frequency in an ultrasound signal, making this an ideal form of radiation to investigate poorly understood features of wave phenomena such as wave propagation under conditions of strong scattering.

1.6 THE ADIABATIC IDEALIZATION

At all frequencies considered in this work, propagation is adiabatic in homogeneous media. This means that, despite the temperature fluctuations which inevitably accompany the pressure fluctuations of sound, thermal dissipation is small and it is the adiabatic compressibility which matters. Care must therefore be taken when comparing ultrasound-determined elastic constants with those determined using static methods. In the static case, it is the isothermal constants which are measured.

Pierce (1981c) shows that the adiabatic approximation holds at the very lowest frequencies of sound propagation. This gives the important result that the limit of frequency tending to zero ($f \to 0$, the adiabatic limit) is not the same as frequency equals zero ($f = 0$). Although this result may appear counterintuitive, it is correct, and anyone unconvinced after reading Pierce can find a separate proof in Zemansky (1957b). Overall, the rate of dissipation of heat falls as wavelength increases. Hence, the lower the frequency, the more accurate the adiabatic approximation. Pierce shows that the adiabatic approximation begins to break down at frequencies above 10^{12} Hz in water and 10^9 Hz in air. The adiabatic approximation also breaks down when applied to inhomogeneous materials. Under these conditions isothermal propagation may occur over small regions at the boundary between two materials, resulting in so-called thermal scattering of the sound wave. This case is dealt with in detail in Chapter 4.

Pierce also shows (1981c) that sound velocity in the adiabatic approximation is virtually independent of frequency. Throughout this work it will be assumed that, in the absence of scattering, this is the case, and frequency will not generally be quoted for sound velocity measurements. When frequency is given, it is because scattering or relaxation are significant and velocity dispersion (frequency dependence) is a possibility.

1.7 COMMON SENSE IS UNSOUND

It is the purpose of this book to show that with existing technology and commercially available apparatus, a very wide range of applications are within reach of the working scientist, technologist, and engineer.

To achieve this objective it is necessary to understand the factors that determine velocity of sound. Here it is appropriate to deal with a few misconceptions about ultrasound metrology. Many people who have worked in laboratories will remember the buzzing of the ultrasound bath with less than fond memories. This has created an impression in the back of the minds of many that ultrasound is a material altering technique, which, indeed, the ultrasound bath is. However, the power levels in ultrasound baths are up to a million times higher than the typical ultrasound velocity instrument. The power levels in the baths must be high enough to create cavitation, which is the process that actually creates material changes. An ultrasound velocity measurement, on the other hand, is usually carried out at much lower power levels and normally leaves the material unchanged. The pulse-echo technique usually used to determine sound velocity causes very small elastic displacements in the material which are well below the cavitation threshold for water. In fact, it is only because water is so susceptible to cavitation that ultrasound baths work as well as they do. If it were not for cavitation, ultrasound propagation in most baths would be nondestructive.

Another consequence of the use of low sound power levels (e.g., below ca. 10 kW m^{-2} in water and at room temperature; Puskar, 1982) is that the displacements are elastic, i.e., strain is linearly related to the stress. This means that the velocity will be independent of the sound power levels, an important assumption made throughout this work.

There is a second very important misconception about ultrasound which is a serious barrier to its widespread use. For one reason or another, many people believe that the velocity of sound depends mainly on the density of the material through which it propagates. A connected misconception is that sound velocity in a mixture of materials will be some average of the velocities in the components of the mixture. Neither of these beliefs is true, yet they are perpetuated even by some manufacturers of ultrasonic instruments, who believe these things themselves. In fact, as will be seen (Chapter 2), sound velocity in liquids depends on both density and compressibility.

The preceding discussion demonstrates the need to place ultrasonic measurement and interpretation of data on a firm theoretical footing.

1.8 SCOPE OF THIS WORK

Since it is much simpler and cheaper to make accurate measurements of sound velocity than it is to measure the absorption of sound accurately, this work concentrates on velocity measurement and the inference of useful data from such measurements. For the reader interested in a more general overview of ultrasound applications, Kuttruff (1991a) is a good introduction. For an in-depth study of the determination of thermodynamic and kinetic information from velocity and absorption of sound data, see Trusler (1991). Matheson (1971) is an introduction to the study of molecules by sound velocity and absorption methods. An introduction to shear wave methods for the determination of viscoelasticity as well as a general overview of the development of the subject of physical acoustics since 1964 can be found in Mason *et al.* (1964–1992). A review of experimental methods can be found in Edmonds (1981). A comprehensive review of broadband techniques is in Eggers and Kaatze (1996).

In this book, it is assumed that the reader will purchase the necessary apparatus for sound velocity measurement from an instrument manufacturer who will have solved the acoustical problems associated with the measurement in question. The central question addressed here is that of data acquisition and interpretation.

A number of assumptions have been adopted throughout this work and will apply unless otherwise stated:

The adiabatic approximation applies.
Sound velocity is independent of frequency.
Sound velocity is independent of power level.
Only longitudinal sound waves propagate over macroscopic distances, i.e.,
  we consider only liquids or materials which approximate to liquids by
  virtue of a small or zero shear modulus.
Ultrasound wavelength is much smaller than the sample dimensions and
  much greater than the size of any objects dispersed within the sample.

1.9 HOW TO USE THIS BOOK

It is assumed that the reader is unfamiliar with ultrasound. Concepts and theory are only introduced when they become necessary for the explication of a given experimental technique. To minimize repetition, the reader will often be referred to other sections to obtain a complete theoretical picture. Fundamental concepts of wave propagation are introduced in the first two chapters and the apparatus of scattering theory is explained in Chapter 4. In order that the reader not be intimidated by equations, some scattering theory results are used in Chapter 3 before being derived later. Where there is not enough space to deal with a topic adequately, the reader is referred to other works. An appendix gives a basic introduction to ultrasound theory and summarizes some of the main results. The index has been compiled to aid understanding. A glossary defines all variables used in equations in the book; there are a few cases where the same symbol has different meanings. Where this is the case, care has been taken to ensure that the different meanings apply in quite different contexts.

2

WATER

Water is an important component of very many liquid and soft-solid materials. In some respects, its acoustical behavior is unlike that of most other liquids. For example, below about 74°C, the temperature coefficient of velocity in water is positive, whereas in most other liquids it is negative. Another unusual property of water which is relevant to its acoustical properties is its negative value for the temperature coefficient of volume expansivity below 4°C. On the other hand, a great deal of high-quality data on the acoustical behavior of water is available and is a good example of what can be achieved with ultrasound measurement techniques. For these reasons, this chapter is devoted to ultrasound propagation in water and dilute mixtures of water with other materials. The example of measurements in water will also be used to illustrate a number of general features of ultrasound velocity measurement and data interpretation.

2.1 MEASUREMENT OF SOUND VELOCITY

2.1.1 INTRODUCTION

The measurement of the velocity of sound is a straightforward affair that can be done quickly and easily. Today's equipment is more sophisticated and certainly more accurate and convenient, but it does not differ in principle from the technique used by Colladon and Sturm (1827) in Lake Geneva. A block diagram of one apparatus used by the author appears in Figure 2.1. This technique is variously known as "pitch and catch," "pulse-echo time of flight" or "Acoustic Time of Flight Measurement" (AToM).

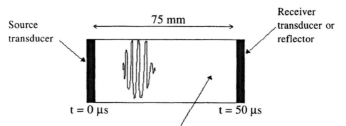

Velocity in sample equals distance / time, for example,
75 mm / 50 μs = 1500 m s^{-1} for single transit and
150 mm / 100 μs = 1500 m s^{-1} for a single reflection

FIGURE 2.1 Block diagram of an acoustic pulse-echo (AToM) apparatus. A representation of the pulse, based on an oscilloscope trace, is shown in the picture and explained in the two figures that follow.

In Figure 2.2 is a time-varying and space-varying signal. The time variation is characterized by the frequency f and the spatial variation by the wavelength λ (Figure 2.3). The two are related through the velocity of sound, which is given by

$$v = f\lambda. \qquad (2.1)$$

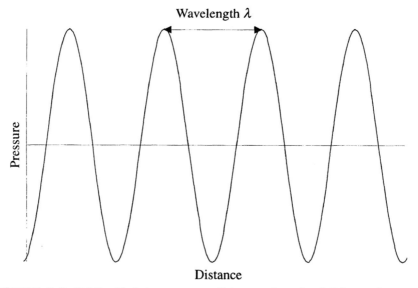

FIGURE 2.2 Relationship between pressure, distance, and wavelength λ for one frequency component of the signal. λ is 1.482 mm in water at 20°C and 1 MHz. In the pulse shown in Figure 2.5, the wavelength varies from each end to the middle because of the range of frequencies which make up the pulse.

2.1 MEASUREMENT OF SOUND VELOCITY

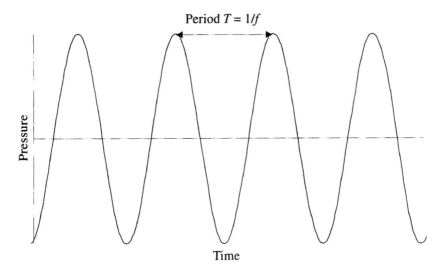

FIGURE 2.3 Relationship between pressure, time (t), period (T) and frequency (f) in one frequency component of the time-varying signal in Figure 2.1. The signal frequency of a pulse in Figure 2.5 will have a center frequency of 1 MHz, but because it is a pulse it will contain a range of frequencies extending from approximately 0.8 to 1.2 MHz.

In Figure 2.4 is illustrated a straightforward implementation of the pulse-echo technique which has been used for much of the data presented in this work. A simple start/stop timer is used to obtain the time of flight. All the circuitry can be

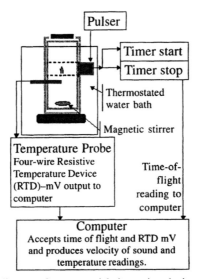

FIGURE 2.4 Block diagram of a commercial ultrasonic velocimeter operating at 2.25 MHz.

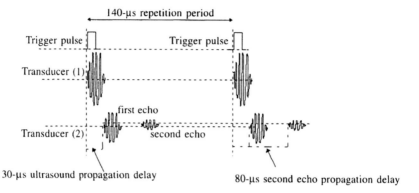

FIGURE 2.5 Pulse waveforms commonly found in an apparatus configuration similar to that in Figure 2.1.

integrated into a single small box which uses serial communications to send data to a computer. Calculation of time-of-flight and temperature is carried out in the computer. A pulse of sound is created by electrically stimulating the source transducer (Figure 2.4), a lead zirconium titanate or other piezoelectric ceramic disc. The excitation pulse also starts a timer. Typically the pulse may last a few microseconds and consist of two or more cycles (Figure 2.5). The pulse travels through the epoxy wear plate which protects the front of the transducer and passes into the liquid filling the sample cell. Reflections at the front face of the transducer are damped out by a tungsten-loaded epoxy backing. Once in the sample the pulse travels across the cell to the far wall, where it is reflected and travels back to the transducer. The returning pulse is detected and stops the timer; the resultant time is then fed to the computer. The timing circuit is precise to within 20 ns. At the same time, the temperature in the sample is measured with a resistive temperature device (RTD), a four-wire platinum resistance thermometer providing an accuracy of 0.1°C and a precision of 0.02°C.

2.1.2 ACCURACY AND ERRORS

With ultrasonic velocity measurement using computer-based equipment, the distinction between "precision" and "accuracy" becomes significant. Precision can be understood by reference to a watch which displays seconds. The precision of this instrument is obviously one second. However, its accuracy will vary, depending on how its time has been set. For example, if it has been set to an electronically transmitted time signal it could well be accurate to one second. However, accuracy will be far lower if the instrument is systematically slow by one minute in every hour. Systematic errors may also be introduced, for example, by incorrect reading of a reference clock. Random errors may be introduced, for example, through ambient temperature variations affecting the timekeeping. The reader is referred to Taylor (1982) for an interesting and comprehensive introduction to the subject of the analysis of errors.

2.1.2.1 Temperature

The temperature is important, because the path length can vary with temperature and because the velocity of sound in water is strongly dependent on temperature. Since the velocity of sound in water varies by approximately 3 m s^{-1}°C^{-1} at 20°C, an error in temperature of 0.1°C will produce an error in the velocity of 0.3 m s^{-1}.

2.1.2.2 Acoustical Delays

Overall accuracy depends to some extent on the care with which the factory removes the acoustical delay due to the probe wear plate (Figure 2.6), whose variable thickness introduces a variable delay into the total transit time of the pulse. This adjustment is carried out by setting a constant delay to the start time of the timer. In an epoxy wear plate of 0.1 mm thickness, a delay of approximately 20 ns will be created. This delay is of the same order as the accuracy of the timing circuitry employed in commercial equipment. Wear plates of greater thickness than this require their delays to be electronically removed in the factory. The systematic errors arising from these delays are not totally removed by the calibration procedure described later.

The composite construction (Figure 2.6) of piezoelectric-ceramic transducers is often a source of unreliability. Differential contraction between the epoxy and the ceramic may result in delamination of the transducer and a gradual failure. This process can generate errors whose origin may initially be difficult to identify. Degradation in the electrical connection between the transducer electrodes and the transducer connector is another common source of problems.

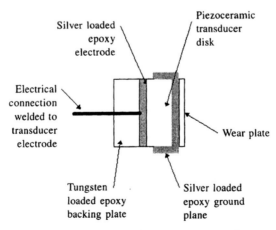

FIGURE 2.6 Constructional elements of an ultrasonic transducer manufactured from piezoceramic material.

TABLE 2.1 Typical Power Levels and Other Propagation Parameters for Ultrasound Propagation in Water at 1 MHz and 30°C [a]

f (MHz)	T_0 (°K)	P_0 (MPa)	I (kW m^{-2})	Δp (MPa)	s ($\times 10^{-6}$)	ξ (nm)	ξ' (mm s^{-1})	ξ'' (km s^{-2})	Z (MPa s m^{-1})	ΔT (mK)	ξ'/v ($\times 10^{-6}$)
1	303	0.1	0.1	0.017	7.6	1.8	11.5	72	1.47	.38	7.6
1	303	0.1	10	0.17	76	18	115	720	1.47	3.8	76
1	303	0.1	1000	1.7	760	180	1150	7200	1.47	38	760

[a] Here f is frequency (in MHz), T_0, absolute temperature (K); P_0, absolute pressure (MPa); I, intensity (kW m^{-2}); Δp, pressure change from P_0 owing to the passage of ultrasound (MPa); s, the condensation ($= [\rho_0 - \rho]/\rho_0$); ρ_0, the static density (kg m^{-3}); ρ, the instantaneous density (kg m^{-3}); ξ, the particle displacement (μm); ξ', the particle velocity (mm s^{-1}); ξ'', the particle acceleration (km s^{-2}); ΔT, the temperature change owing to the passage of the ultrasound (K); Z ($= \Delta P/\xi' = \rho v_l$), the specific acoustical impedance (Pa s m^{-1}); and v the velocity of a compression ultrasound wave. From Povey and McClements (1989).

2.1.2.3 Impedance

The *electrical impedance* is defined as the ratio of the voltage to current, and the *acoustical impedance* is defined as the ratio of the acoustic excess pressure Δp_0 to the particle velocity ξ' (see Table 2.1). In general the impedance (Z) is a complex quantity, comprising a resistive ($R_z = \Re e(Z)$) and a reactive component ($X_z = \Im m(Z)$). In the acoustic case,

$$Z = \frac{\Delta p}{\xi'} = \rho \frac{\omega}{k} \approx \rho v. \qquad (2.2)$$

The quantity ρv is often called the characteristic impedance. Matching the impedance of a transducer to both the electrical circuitry and the sample is important to the success of any experiment, because if this is not achieved, then power transfer from energy in an electrical form will not be transmitted into the sample. This is because an impedance *mismatch* causes reflection of the wave. To understand this better, consider that a pane of glass is very well matched to the free space through which light travels. So most of the light travels through the glass. But if the glass is silvered, most of the light will be reflected. Similar processes occur in acoustics (see Figure 2.7).

The ratio of the acoustic *amplitude* (displacement in this case) incident on the interface to that within the interface is called the *transmission coefficient* T. The ratio of the reflected amplitude to the incident amplitude is called the *reflection coefficient* $\Re$. They are given by

$$T = \frac{\xi_t}{\xi_i} = \frac{2Z_1}{Z_1 + Z_2} \qquad (2.3)$$

2.1 MEASUREMENT OF SOUND VELOCITY

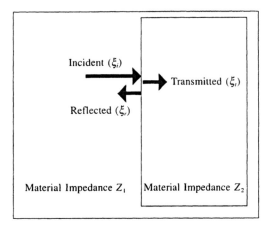

FIGURE 2.7 Reflection and transmission of ultrasound through materials depends on their relative acoustic impedance.

$$\Re = \frac{\xi_r}{\xi_i} = \frac{Z_1 - Z_2}{Z_1 + Z_2}. \tag{2.4}$$

Manufacturers offer transducers matched either to steel or to water, and a choice must be made to achieve the best match. *"Reverberation"* is associated with acoustical mismatching. It is of concern wherever different materials form part of an acoustical path, for example, if transducers are coupled through water into the glass walls of a cuvette containing the sample. If the path length of an obstacle causing reverberation is small enough (Figure 2.8) then each reverberation will add a phase shift to the pulse, creating complicated changes in the pulse envelope.

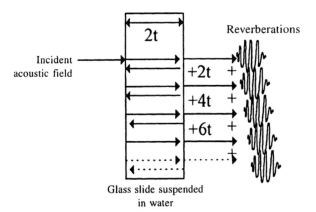

FIGURE 2.8 Partial reflection of an acoustical pulse within a material obstacle (e.g., a glass sample wall) of acoustical time delay t. Each reverberation is displaced slightly relative to the next. When all the reverberations are summed in the receiver transducer, the resultant waveform displays interference between successive reverberations.

2.1.2.4 The Control of Reverberation with Buffer Rods

One method of minimizing the effects of reverberation is to employ buffer rods whose length and diameter are much greater than the acoustic wavelength. The buffer rod needs smooth faces manufactured to a high degree of parallelism. The criteria for both parallelism and roughness are related to the need to minimize the effects of phase variation across the rod. If amplitude is required to be accurate to within 1% (b in Figure 2.9 is 100) and the acoustic field variation is approximated as a triangular wave, then the phase variation needs to be less than $\frac{1}{400}$ of a wavelength.

This phase variation, at 1 MHz in quartz (velocity = 5.9×10^3 m s^{-1}; Weast, 1988b), is equivalent to a distance of 15 μm ($V = f\lambda$, $\lambda = 5.9 \times 10^{-3}$, $\lambda/400 \approx 1.5$ μm). The parallelism required between the two faces of the buffer rod in order to achieve the aforementioned criterion is $2 \times$ length of the rod $\times$ angle = 15 μm. Thus, the faces must be parallel to within $15 \times 10^{-6}/80 \times 10^{-3} \approx 2 \times 10^{-4}$ rad for a 40 mm buffer rod. This is equivalent to a variation of less than 2 μm across a 10 mm diameter buffer rod face. Similar strict criteria apply to velocity measurement. The material from which the rod is made also requires a high level of uniformity; fused quartz meets this criterion admirably. A two-buffer rod configuration is illustrated in Figure 2.10.

Reverberation can be exploited in a subtle way in a device called the *quarter wave transformer*. If an obstacle has a thickness exactly equal to one quarter-wavelength of sound within it, then it will match the impedance on each side, effectively transmitting the sound. If the thickness is half the wavelength of sound, then the obstacle becomes a perfect mirror. This is the basis for optical coatings on glass lenses which screen out, for example, ultraviolet and blue light. The same principle works for sound. The difficulty is that most transducers do not produce a pure enough frequency for the technique to be really effective.

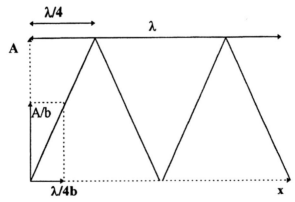

FIGURE 2.9 Example calculation of the accuracy required in position, x, to obtain an accuracy in amplitude of A/b, where A is the maximum amplitude. The sinusoidal variation is approximated by a triangular waveform.

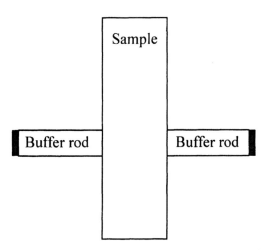

FIGURE 2.10   Two buffer rods in a transmitter/receiver configuration.

2.1.2.5 Acoustical Bonds

Care needs to be taken in the process of acoustically bonding the transducer, buffer rod and sample. Bonds may contribute errors and variability in the acoustical path because they normally have a finite thickness. These errors may arise in all the ways explained in §2.1.2. For compressional waves, viscous liquids generally make the best bonds because they form a uniform and thin bond between the two surfaces, effectively excluding air. Care needs to be taken to ensure that air is not trapped within the bond. Depending on the application, the bonding material may be water soluble or water insoluble. Although permanent bonds formed with materials such as araldite and Super Glue can be very effective, any problems with the bond can result in the destruction of the bonded components. For this reason, materials such as "Nonaq" stopcock grease (Fischer Scientific, USA), low-viscosity silicon grease, high-viscosity silicon oil, proprietary water-soluble pharmaceutical jelly, and various proprietary acoustic bonding gels are widely used as bonding materials.

2.1.2.6 Power Levels

In a liquid, the sound pulse propagates as alternative increases and decreases of the pressure in the material, which together move through the material at the velocity of sound. Typical equivalent electrical power levels at the transducer vary between 20 volts and 1 kV delivered into an impedance whose magnitude is approximately 70 Ω. This is a complex impedance, comprising both resistive and inductive components, which vary from transducer to transducer. It is an electromechanical quantity, so the electrical impedance is affected by the acoustic impedance. It is also very frequency dependent. With conversion efficiencies of 20% or so, instantaneous power levels at the transducer face will be a maximum of 50 watts or so, and with an aperture diameter of the order of 400 mm², we expect an acoustical power flux of 12.50 kW m⁻². This translates into particle displacement amplitudes of 20 nm in water (see Table 2.1) and temperature changes of a few millikelvins).

The associated pressure levels can also be very high, of the order of hundreds of kilopascals (Table 2.1). Thus, the peak instantaneous power level at the face of a transducer driven by a 1 kV pulse is theoretically capable of generating cavitation in water (Table 2.1). The average power levels in the pulse-echo experiment are very much lower than this, however. A typical repetition rate for the pulse is 500 Hz, so a three-cycle, 1-MHz pulse will last for 3 µs and will repeat every 2 ms. The ratio of on-time to off-time is called the *duty cycle* and is about 2×10^{-3}. The mean square power in the pulse is approximately the peak power divided by $\sqrt{2}$, and when this is multiplied by the duty cycle we obtain the average power levels in a pulse-echo experiment of the order of watts, i.e., hundreds of times lower than that required to produce cavitation. The pulse nature of the signal also makes cavitation less likely, since the cavitation bubble requires several cycles of ultrasound to grow, in a process called *rectified diffusion* (Apfel, 1981).

Examination of Table 2.1 shows that although the linear approximation ($\xi'/v_1 \ll 0.1$; $\Delta T \ll 100$ mK) is valid over a very wide range of power levels, local pressure variations can easily reach 1 atm at higher power levels and particle accelerations can be large. Under these conditions cavitation, to which water is particularly prone, becomes likely so that the ultrasonic technique may no longer be described as nondestructive. This is the basis for the figure of 10 kW m⁻² quoted earlier for the cavitation threshold for water. Perhaps surprisingly, these power levels are much more likely to be reached with medical ultrasound apparatus than with the type of equipment used for ultrasound velocity measurement.

We can conclude therefore that a pulse-echo experiment is nondestructive under most conditions, and furthermore that information is available 500 times a second or more. It is therefore appropriate to consider ultrasound for monitoring rapidly changing processes.

2.1.2.7 Diffraction and Phase Cancellation

The transducers considered in this work are all piston sources of sound (see Chapter 1). This means that they may be described mathematically as if they were simply moving forward and backward in a manner similar to that of a piston. The wavefront resulting from this type of source is not a straightforward plane wave (Figure 2.11). It is therefore necessary to consider briefly the implications of the use of piston-type sources.

Although a piston source will generate a wavefront of nearly equal phase across its face (Figure 2.12 and Figure 1.3), a signal which has passed through a sample need not maintain the phase relationships across its wavefront. For example, the wavefront that leaves a piston source is accurately described as plane (flat) within a distance r^2/λ, where r is the transducer radius. The cylindrical space, r^2/λ long and $2r$ in diameter, in front of the transducer is called the *near field*. In this region

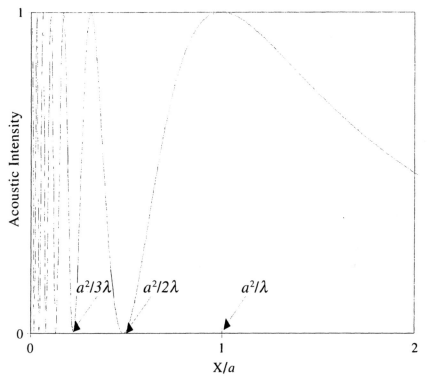

FIGURE 2.11 The axial intensity as a function of distance in front of a piston transducer operating with diameter $2a \gg \lambda$. This figure has been plotted using Equation 5-7.3 in Pierce (1981d).

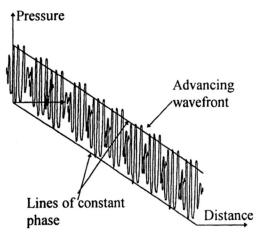

FIGURE 2.12 Instantaneous pressure versus distance plot for an ultrasound pulse advancing on a plane wavefront.

the signal is diffracted according to *Fresnel diffraction* (Figure 2.11). Although the wavefront is nearly plane within the near field, the amplitude varies wildly with distance and so the near field is generally unsuitable for amplitude measurements because the signal amplitude is so sensitive to the distance from the transducer. However, so long as the amplitude is large enough to detect, time of flight measurements can be made within the near field (Breazeale *et al.*, 1981); the contribution to the error from phase variations of diffraction origin is of the order of 0.005%.

Beyond the near field lies the *far field*. In this region, signal amplitude is an exponentially decaying function of distance (see Figure 2.11) and the wavefront is no longer plane. Instead it spreads out, tending towards spherical. The diffraction of the wavefront in the far field contributes roughly one decibel of attenuation per distance of r^2/λ (Seki *et al.*, 1956) and is called *Fraunhofer diffraction* (Figure 2.13). Diffraction of the wave can also introduce timing errors by generating a variation of phase across the transducer face. Phase variations can also be introduced by nonparallelism between the generator-transducer and the receiver-transducer or reflector, and these variations generally increase with increasing acoustic path length.

Strong multiple scattering of a wavefront can reduce or destroy its coherence (Figure 2.14), so that it becomes partially or completely undetectable by phase-sensitive transducers.

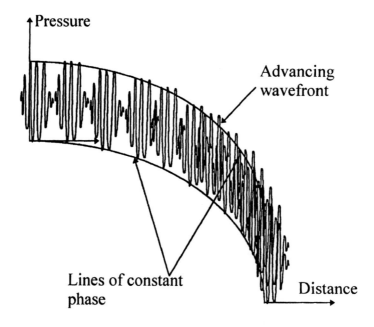

FIGURE 2.13 The wavefront of Figure 2.12, undergoing Fraunhofer diffraction.

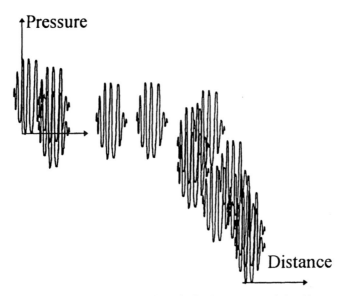

FIGURE 2.14   The wavefront of Figure 2.12 with its phase relationships destroyed.

2.1.2.8 Timing Errors Due to Trigger Point Variation

Figure 2.5 illustrates a problem associated with the pulse echo technique, which may not be apparent to the inexperienced user. The figure illustrates the pulse waveforms detected by an oscilloscope connected to both the transmitter (Transducer 1) and receiver (Transducer 2). This is consistent with the apparatus of Figure 2.1 being operated with a receiver on the right-hand side, not a reflector. The uppermost trace is the trigger pulse, which is used to synchronize timing in the electronic detection system (Figure 2.4). The lower trace represents the signal which has passed through the liquid sample and has been attenuated. A second echo can also be seen, this has traversed the cell width three times, in contrast to the single crossing of the cell experienced by the first "echo." Commercial instrumentation normally measures the time between the trigger pulse and the first "echo." This contains delays associated with the transducer wear plate and the electronics which have been referred to above.

Examination of Figure 2.15 shows that variation in signal amplitude, when the signal is just big enough to trigger, can produce timing errors of up to several times the period of the wave. For a 1-MHz signal this could mean an error of several microseconds in a 50-microsecond transit time, or an error of nearly 10%. Clearly this is unacceptable. This situation is characterized by sudden jumps in the transit time when the peak amplitude changes slightly and is normally not difficult to spot with computer logging systems. However, it is difficult to see in commercial ultrasound velocimeters which dispense with oscilloscopes to reduce costs, and which are operated manually.

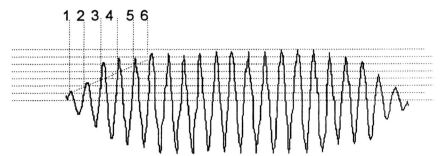

FIGURE 2.15 The effect of trigger level on the position of the trigger point for a 19-cycle tone burst applied to a medium damped crystal.

2.1.2.9 Measuring Group Velocity

The pulse-echo method measures the group velocity of a wave; this is the velocity with which the pulse envelope travels. Commercial instrumentation usually detects the first negative-going part of the incoming waveform and assumes that this is characteristic of the whole envelope. (See Section 2.1.1.) The group velocity is defined as

$$v = \frac{d\omega}{dk}, \qquad (2.5)$$

where

$$k = \frac{2\pi}{\lambda} \qquad (2.6)$$

is called the wave number, λ is the wavelength, and ω is the radial frequency $\omega = 2\pi f$; f is the frequency.

A second velocity, called the phase velocity (v_p) is defined as follows:

$$v_p = \frac{\omega}{k}. \qquad (2.7)$$

This is the velocity of a wave of a single frequency; since this cannot exist in isolation as a pulse, the phase velocity cannot be measured directly with pulse techniques. It can be measured indirectly and requires a broad-band ultrasonic system of greater sophistication than the simple instrumentation described thus far. A block diagram of a system capable of measuring both signal amplitude and phase, and hence of determining phase velocity and attenuation, is shown in Figure 2.16; this will be discussed in more detail in the chapter on particle sizing.

In the case of water the distinction between the various definitions of velocity may be ignored under most circumstances. The instrumentation discussed earlier in Figure 2.4 is only appropriate for use in samples where the distinction between the two definitions of velocity is not significant.

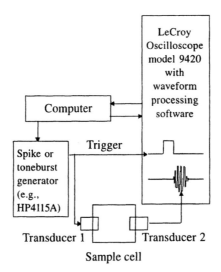

FIGURE 2.16 Broadband ultrasound detection system.

2.1.3 CALIBRATION

Many of the errors associated with this type of measurement can be eliminated or reduced by calibrating with double-distilled water, at the required temperature. The calibration is carried out to get a very accurate determination of the path length.

This calibration should be performed at all operating temperatures for maximum accuracy and the data of del Grosso and Mader (1972) used as the water standard. This method is very strongly recommended because it enables good accuracy to be obtained with relatively low-cost equipment.

Experience suggests that the pulse-echo technique applied to liquids and using a calibration technique is capable of an accuracy of 1 m s^{-1} and a precision about 10 times better. However, if high sample attenuation is experienced, serious errors can occur.

2.2 THE DEPENDENCE OF VELOCITY OF SOUND ON DENSITY AND COMPRESSIBILITY

How does the velocity of sound depend on the properties of the material through which it propagates? A treatment of this subject whose sophistication is still contemporary was given by Lord Rayleigh in 1877 (Strutt, 1877). For our purposes, so sophisticated a treatment is inappropriate to begin with. A more convenient place to start is the textbook written by Wood in 1941 (Wood, 1941a, 1964). Despite its age, this is still the best starting point for those interested in the subject.

Wood starts from the view that the physical properties of the medium which influence sound velocity are density and elasticity, corresponding to the mass and stiffness of a particle vibrating in a sound field. As the wave of compression and rarefaction passes through the material, the volume and density fluctuate locally about the normal values. Thus, the important quantities are *dilation* ($\Delta = \delta V/V_0$), where V is the *instantaneous volume* of a local volume (defined as a vanishingly small volume element of material) and V_0 is the *original volume*; *condensation* ($s = \delta\rho/\rho_0$) where ρ is the *instantaneous density* and ρ_0 is the *original density*. For the very small displacements which occur in our experiments $s = -\Delta$ to a very good approximation. This is called the linear region. The *bulk modulus of elasticity*, *bulk modulus* for short, is then $B = -\delta p/\Delta = \delta p/s$, where p is the *instantaneous pressure*, δp the *stress*, and δV the corresponding *strain*. That is, the adiabatic bulk modulus is the ratio of stress to strain, multiplied by the volume of material being strained and has units of pressure (Pa). The *adiabatic compressibility*, κ, of a pure liquid is then the reciprocal of the adiabatic bulk modulus, B, $\kappa = 1/B$. Wood shows that the velocity of sound in homogeneous liquids and gases is independent of frequency and given by the following equation, which is named after him:

$$v = \sqrt{\frac{B}{\rho}} = \sqrt{\frac{1}{\kappa\rho}}, \qquad (2.8)$$

where v (m s^{-1}) is the velocity of a sound wave in which the particle displacement is parallel to the direction of propagation. This equation is called the *Laplace equation* by physical chemists (Nomura *et al.*, 1987) and the Newton–Laplace equation by a few others. The *Wood equation* (Equation 2.8) indicates that the velocity of sound is *solely determined by the elasticity and density of the material through which it is passing*. We will see later that sound propagation is in general more complicated than this.

2.2.1 THE VELOCITY OF SOUND IN MIXTURES AND SUSPENSIONS

According to Wood (1941), it is the mass and the stiffness of the material through which the sound passes which are the crucial factors determining sound velocity. This is because the propagation of sound is described very well by Newton's second law, whereby a force acting on an element of the material accelerates the material. Since the displacements are small, they may be assumed to be *harmonic* (any oscillatory motion in which the restoring force is proportional to displacement may be called harmonic). This is called the linear approximation. The magnitude of the fluctuations in volume (dilatation) and density (condensation) associated with the sound wave is controlled by the properties of the medium and the applied forces. The velocity of sound in mixtures and suspensions will therefore be controlled by the *mean density* and the *mean compressibility*. This fundamental

2.2 DEPENDENCE OF VELOCITY OF SOUND ON DENSITY AND COMPRESSIBILITY

observation appears in Wood (1941b), who applied it to fluid mixtures. However, its systematic development is attributed to Urick (1947), who generalised it to dispersions of solid particles, and the resulting equation is now called the *Urick equation*:

$$v = \frac{1}{\sqrt{\kappa \rho}}, \quad \kappa = \sum_j \phi_j \kappa_j, \quad \rho = \sum_j \phi_j \rho_j. \tag{2.9}$$

v is the sound velocity in the dispersion, and ϕ is the volume fraction of dispersed phase. In the case of one material dispersed within another, the compressibility and density may be written

$$\kappa = \phi \kappa_2 + (1-\phi)\kappa_1, \quad \rho = \phi \rho_2 + (1-\phi)\rho_1. \tag{2.10}$$

Here, the subscripts refer to the constituent phases; in particular 1 refers to the continuous phase, 2 to the dispersed phase in a two-phase system. (In a three phase system, 2 refers to the liquid dispersed phase and 3 to the solid dispersed phase.)

If Equation 2.9 is to be used to determine the volume of material in a two-phase system, it can be rewritten in a more convenient form:

$$\phi = \frac{-B \pm \sqrt{B^2 - 4AC}}{2A}, \tag{2.11}$$

where

$$A = v_1^2\left(1 - \frac{\rho_1}{\rho_2}\right) + v_2^2\left(1 - \frac{\rho_2}{\rho_1}\right)$$

$$B = v_2^2\left(\frac{\rho_2}{\rho_1} - 2\right) + v_1^2 \frac{\rho_1}{\rho_2}$$

$$C = v_2^2\left(1 - \frac{v_1^2}{v^2}\right).$$

From Equation 2.11 we can conclude that, in order to determine the unknown volume of material in a mixture of two materials, it is necessary, in addition to determining the velocity of sound in the mixture, to know the density and sound velocity of the constituent components.

The volume fraction, ϕ, can be converted into a mass fraction, w:

$$w = \frac{\phi \rho_2}{\phi \rho_2 + (1-\phi)\rho_1}. \tag{2.12}$$

In Chapter 3, this method of determining an unknown volume of material in a mixture will be considered in detail.

The consequences of Equation 2.9 can be startling. First, it is a quadratic with two roots (Equation 2.11). Wood (1941b) gives the example of a determination of the velocity of sound in a mixture of nonresonant air bubbles and sea water. In Figure 2.17, Wood's figures have been recalculated using data for distilled water, which are more appropriate in this work.

2.2.2 THE VELOCITY OF SOUND IN AIR/WATER MIXTURES

The following features of Figure 2.17 are of great importance when considering the impact of suspended air bubbles on acoustic propagation in water. First, regardless of resonance, very small amounts of undissolved air can have a dramatic

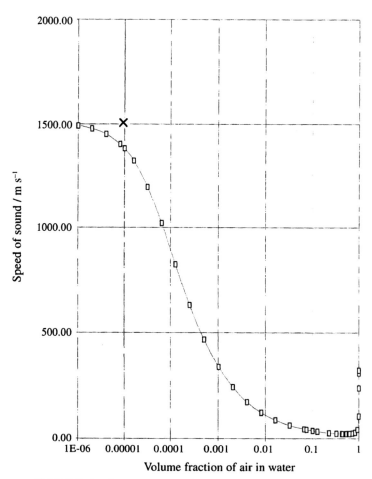

FIGURE 2.17 The velocity of sound in air/water mixtures at 20°C.

2.2 DEPENDENCE OF VELOCITY OF SOUND ON DENSITY AND COMPRESSIBILITY

effect on the velocity of sound. In Figure 2.17, the ✕ indicates the velocity of sound in the absence of air, so 10 ppm of undissolved air changes the velocity by 8% and 0.1 ppm (not shown on the figure) changes the velocity by 1.4 m s^{-1}, an easily measurable amount. Second, the velocity reaches a minimum of 22.27 m s^{-1} at 53% undissolved air. This is far smaller than the velocity either in water (1505.76 m s^{-1}) or in air (322.16 m s^{-1}). Third, the air content has to exceed 90% before the velocity increases rapidly towards its pure value.

Finally, in general there will be two solutions for each measured velocity. This is not necessarily always the case, but additional information is often required in order to distinguish the correct solution in any given set of circumstances.

The features exhibited by the concentration dependence of sound speed in the example of the air/water mixture shown earlier, are present in all mixtures to a greater or lesser extent.

2.2.3 THE IMPORTANCE OF REMOVING AIR FROM SAMPLES

Although dissolved air (as opposed to air bubbles) has a very small effect on the velocity of sound, small changes in pressure can bring dissolved air out of solution, creating bubbles. Moreover, naturally occurring waters often contain tiny bubbles, so-called microbubbles, whose size can be of the order of 1 µm or less. Therefore, it is essential to take great care to de-aerate aqueous systems, prior to acoustic measurement. This is may be done by sonication—exposing the sample to continuous irradiation with high-power ultrasound. This is quickly done by immersion in an ultrasonic cleaning bath. For this to be successful, it is necessary for the continuum to have a low viscosity so that the bubbles can cream out of the sample. If this is not the case, it may be necessary to bubble nitrogen through the sample. The nitrogen displaces air from solution and removes it from the sample. Nitrogen itself is not so soluble and helps remove the bubbles.

There are circumstances where these de-aeration techniques may fail—in highly viscous materials and in the presence of surfactants which stabilize small bubbles. Under these circumstances it may be necessary to centrifuge the sample. Fortunately, ultrasound is its own excellent detector of gas bubbles (see Chapters 4 and 5), and their presence should be suspected whenever anomalous ultrasound velocity readings are obtained.

2.2.4 THE EFFECTS OF TEMPERATURE ON PROPAGATION IN WATER

It is important to be aware of the temperature dependence of the propagation of sound in water. At temperatures up to 74°C, the temperature coefficient of the velocity of sound in pure water is positive, unlike most other liquids. This complicates the compositional dependence of the velocity of sound in materials containing water. Sometimes these properties can be used to advantage, as will be seen later.

The speed of sound in pure water is shown in Figure 2.18. This is based on the algorithm given in the paper by del Grosso and Mader (1972). It is accurate to within 0.015 m s^{-1}.

The accuracy of this data and the ready availability of pure water provides a simple method of calibrating ultrasound apparatus for velocity of sound measurement. To obtain velocity measurements accurate to within 1 m s^{-1}, it is a

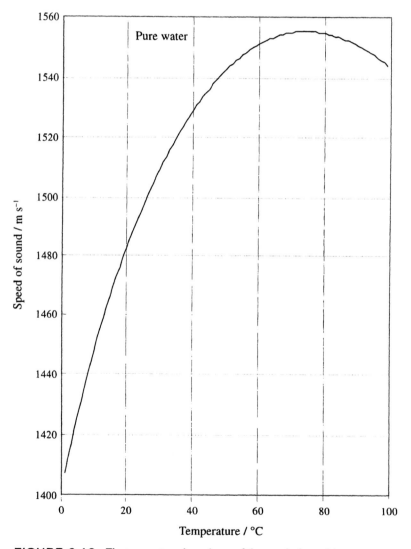

FIGURE 2.18 The temperature dependence of the speed of sound in pure water.

simple matter to measure time of flight to within 10 ns, and it is only necessary then to measure temperature to within 0.3°C at room temperatures.

Take as an example the apparatus depicted in Figure 2.1. The velocity of sound in pure water at 20°C is 1482.336 m s^{-1}. A rule of thumb is that 10 ns is equivalent to 0.3 m s^{-1} over a 74-mm acoustic path, the temperature coefficient of velocity of sound is ~3 m s^{-1}, and therefore ~0.1°C measurement accuracy in temperature is needed to match the timing accuracy.

Data on velocity of sound in pure water can be used to calibrate the acoustic path length, which can then be used as the calibration constant in a velocity of sound apparatus. This method has the virtue of calibrating the entire measurement chain and provides a simple and reliable method of measuring velocity of sound. At temperatures approaching 70°C, the necessary accuracy for temperature measurement is reduced, and at 74°C, an accuracy of 1°C is all that is necessary to give better than 1 m s^{-1} accuracy in velocity. Overall accuracy obtainable in pulse-echo apparatus is of the order of 1 m s^{-1} for velocities of the order of that of water, matched by a precision of the order of 0.1 m s^{-1}.

2.2.5 THE EFFECTS OF PRESSURE ON PROPAGATION IN WATER

Data on sound velocity in distilled water is summarized in Pierce (1981b). Pierce gives an equation for the temperature and pressure dependence of sound velocity between 0 and 20°C and at pressures between 1 and 100 atm (10^5 and 10^7 Pa). The following is an approximate equation which employs del Grosso and Mader's equation (del Grosso and Mader, 1972), modified by the addition of the pressure dependence term quoted by Pierce:

$$v = 1402.39 + 5.03711T - 0.0580852T^2 + 3.33420 \times 10^{-4} T^3 \\ - 1.47800 \times 10^{-6} T^4 + 3.14643 \times 10^{-9} T^5 + 1.6 \times 10^{-6} \left(p_0 - 10^5\right), \quad (2.13)$$

where v is velocity of sound in m s^{-1}; T is temperature in °C ; and p_0 is absolute pressure in pascals.

This equation is accurate to within 0.015 m s^{-1} between 0.001 and 95.126°C and when $p_0 = 10^5$ Pa. It is based on data taken between 0 and 99.9°C. It is likely to be much less accurate under other conditions.

Using Pierce's data for the pressure dependence of density in water together with the Wood equation (Equation 2.8), we obtain the plot of compressibility for distilled water as shown in Figure 2.19.

Accurate data on the pressure dependence of the speed of sound between 0°C and 100°C and between 0 Pa and 8×10^7 Pa can be found in Wilson (1959).

2.2.6 SOUND VELOCITY IN EQUIDENSITY DISPERSIONS

The Urick equation can be rewritten in a more convenient form (Miles *et al.*, 1985), which for an *n*-component system is:

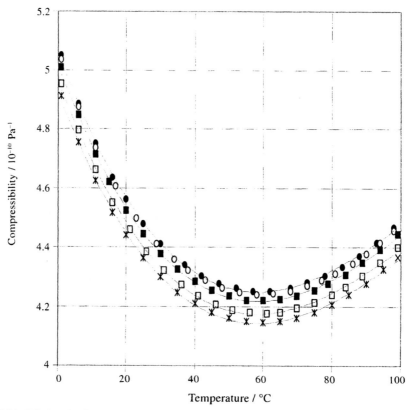

FIGURE 2.19 The pressure dependence of compressibility in distilled water, computed from Equation 2.13. ●, 1.10^5 Pa; ○, 2.10^7 Pa; ■, 4.10^7 Pa; □, 6.10^7 Pa; X, 8.10^7 Pa.

$$\frac{1}{v^2} = \left(\sum_{i=1}^{n} \frac{\phi_i}{v_i^2 \rho_i}\right)\left(\sum_{i=1}^{n} \phi_i \rho_i\right), \tag{2.14}$$

where i represents the ith component and

$$\sum_{i=1}^{n} \phi_i = 1.$$

In many emulsions and dispersions, the density of the suspended or dispersed phase is similar to that of the continuous phase. This is often by design, since equalizing the densities of the two phases will reduce the destabilizing gravitational force of displacement. Under these conditions, the density cancels out of Equation 2.14, which can be written in the following simple form:

$$\frac{1}{v^2} = \sum_{i=1}^{n} \frac{\phi_i}{v_i^2}. \tag{2.15}$$

This equation is also useful for mixtures of materials of similar density, such as for many oils. In fact, the difference between Equations 2.14 and 2.15 is less than 0.2 m s^{-1} for a wide range of oils (McClements, 1988).

2.3 THE RELATIONSHIP BETWEEN VELOCITY AND ATTENUATION—CONDITIONS OF HIGH ATTENUATION

In §2.1.2.9 the phase velocity (v_p) and group velocity (v) of sound are given in terms of the wave vector k. In the presence of the dissipation of energy from the sound wave, **k** is a complex quantity.

$$\mathbf{k} = k' + k'', \tag{2.16}$$

where

$$k' = \frac{\omega}{v_p},$$

and

$$k'' = \alpha.$$

Here α is the attenuation of the sound wave (Neper m^{-1}). We may write an equation for the variation of sound pressure with distance x and time t, in a plane acoustic wave in the form

$$p = p_0 \exp(i(\omega t - \mathbf{k}x)), \tag{2.17}$$

where p_0 is the maximum pressure at the transducer face.

Substituting for k in Equation 2.17 from Equation 2.16, we get

$$p = p_0 \exp(i(\omega t - k'x)) \exp(-\alpha x), \tag{2.18}$$

which is the equation of a pressure wave varying in time with a frequency ω and varying spatially with a wavelength λ; superimposed on top of this variation is an overall decay in amplitude given by the attenuation exponent αx.

Equations 2.5 and 2.6 must be reconsidered in the light of signal attenuation. In particular, the quantity **k** in these equations (see Appendix) is complex, so that

$$\frac{\omega}{\mathbf{k}} = \sqrt{\frac{\mathbf{M}}{\rho}}. \tag{2.19}$$

So long as $\alpha \ll k'$, the effects of attenuation on sound velocity may be ignored, and this is the case unless otherwise stated in this work. For example, at 1 MHz and 20°C in water, $k' = 4245$ m^{-1} and $\alpha = 0.032$ m^{-1}. A calculation of the effects of attenuation on group velocity requires data on the relationship between the wave vector **k** and frequency ω, which depend on the attenuation mechanisms present in the system of interest. In very highly attenuating conditions the concept of group velocity becomes meaningless, sound propagation occurs by a process of diffusion, and the attribute of a wavefront ceases to be meaningful, for example, under conditions of strong multiple scattering such as may be found in bubbly water (see Figure 2.14).

2.4 THE COMPRESSIBILITY OF SOLUTE MOLECULES

2.4.1 INTRODUCTION

The equations for ultrasound velocity in mixtures include the compressibility (Equation 2.9). Hence, measurements of velocity can be used to determine the compressibilities of mixed or dispersed components. This thinking lay behind the development of the Urick equation (Urick, 1947). However, the Urick equation has not been applied in a consistent way to the determination of the compressibility of the dissolved components of solutions. The reasons for this are examined in this section.

Ultrasound velocity measurement, allied with accurate density measurement, is a widely used and indispensable method for determining the adiabatic compressibility of liquids, solutions, solutes, dispersions, and macromolecules (Nölting, 1995; Gekko and Yamagami, 1991; Kharakoz and Sarvazyan, 1993; Gavish et al., 1983). It is also used for the determination of a wide range of thermodynamic parameters (Ewing, 1993). In physical chemistry and biochemistry, measurements of sound velocity and density are converted to compressibility through the Wood equation (Equation 2.8) (usually called the Laplace equation by physical chemists). This procedure is not straightforward and is inconsistent with the Urick equation approach adopted elsewhere in this book.

When a solute is introduced into a solvent, it usually reacts in some way, resulting in changes in its molar volume (due to hydration, for example) and other properties such as specific heat (Larkin, 1975) and compressibility. Since the definition of compressibility involves volume (Equation 2.32), it is essential to account for changes in molar volume in a compressibility determination. Changes in molar volume can be very large; see, for example Figure 2.20.

2.4.1.1 Empirical and Semiempirical Methods

A number of empirical and semiempirical mixing rules have been devised to explain the variation of the velocity of sound with composition. The best known of these are Rao's molecular sound velocity (Rao, 1940) and the Nomoto relation (Nomoto and Kishimoto, 1957; Pandey et al., 1977). Some of these rules have been based on the Urick equation but better fits have sometimes been obtained by empirical methods and there has been considerable debate about the subject (Bonnet and Tavlarides, 1987; Pal, 1994a,b; Tsouris and Tavlarides, 1994). Two issues, emulsion inversion and the efficacy of the Urick equation, are considered later in this chapter and in Chapter 4.

2.4 THE COMPRESSIBILITY OF SOLUTE MOLECULES

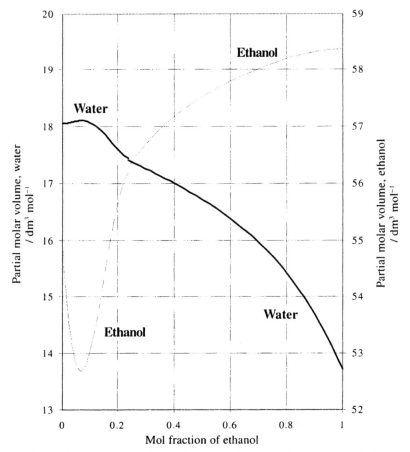

FIGURE 2.20 Partial molar volumes of ethanol and water in binary mixtures of ethanol and water at 20°C. Calculated from density data in Weast (1988a). The breaks in each curve coincide with the changeover from one polynomial to another.

2.4.1.2 Concentrations

Concentration may be expressed in various ways. Up to this point, concentration has been expressed as the volume fraction (ϕ) occupied by a solute. It is more usual when dealing with chemical quantities to express concentration in terms of molarity or molality. This is because an important property of one mole of any pure substance is that it contains 6.022×10^{23} elementary entities, this is called Avogadro's principle. One mole of a substance is its molecular weight expressed in grams.

The molarity of a solution is the number of moles of solute dissolved in 1 liter of solution. The molality of a solution is the number of moles of solute dissolved in 1 kg of solvent. Thus, molality is independent of temperature, whereas the molarity will vary with temperature because of the varying densities of the

TABLE 2.2 Conversion Formulae for Concentration Expressed by Mass[a]

Concentration of solute required	Concentration of solute given				
	w	c_M	m	M	c_2
w		$\dfrac{c_M M_2}{c_M M_2 + (1-c_M)M_1}$	$\dfrac{mM_2}{1000+mM_2}$	$\dfrac{MM_2}{\rho}$	$\dfrac{c_2}{\rho}$
c_M	$\dfrac{\dfrac{w}{M_2}}{\dfrac{w}{M_2}+\dfrac{1-w}{M_1}}$		$\dfrac{M_1 m}{M_1 m + 1000}$	$\dfrac{M_1 M}{M(M_1-M_2)+\rho}$	$\dfrac{M_1 c_2}{c_2(M_1-M_2)+\rho M_2}$
m	$\dfrac{1000w}{M_2(1-w)}$	$\dfrac{1000 c_M}{M_1 - c_M M_1}$		$\dfrac{1000 M}{\rho - MM_2}$	$\dfrac{1000 c_2}{M_2(\rho - c_2)}$
M	$\dfrac{\rho w}{M_2}$	$\dfrac{\rho c_M}{c_M M_2 + (1-c_M)M_1}$	$\dfrac{\rho m}{1000 + M_2 m}$		$\dfrac{c_2}{M_2}$
c_2	$w\rho$	$\dfrac{\rho c_M M_2}{c_M M_2 + (1-c_m)M_1}$	$\dfrac{\rho m M_2}{1000 + mM_2}$	MM_2	

[a] c_2, kg of solute per m³ of solution; c_M, mol fraction; m, molality—gm-mol of solute per kg of water; M, molarity—gm-mole of solute per liter of solution; M_1, molecular weight of solvent; M_2, molecular weight of solute; w, mass fraction of solute; ρ, density of solution/kg m⁻³.

components of the mixture. Concentration is also often expressed in weight of solute per volume of solution or weight of solute per weight of solution. The relationships between the various methods of expressing concentrations is shown in Table 2.2.

2.4.2 DETERMINING PARTIAL VOLUMES

2.4.2.1 The Method of Intercepts

The partial volume, V_{M_2}, of a solute is obtained from the change in total volume of the solution (V) with respect to concentration or molality (Atkins, 1982). It is the change in volume of the solution resulting from a unit change in the concentration of the solute. Accurate determination of the partial volume is vital to the correct determination of solute compressibility, so the procedure for obtaining it will be examined in some detail here. Atkins (1982) gives an account of the method of intercepts which can be used to obtain the partial molar volume. The method of intercepts may be implemented using a spreadsheet and multiple linear regression. The following steps were followed to obtain Figure 2.20.

1. Density data expressed as density against mass fraction of ethanol in water at 20°C was obtained from the *CRC Handbook* (Weast, 1988a). This data may also be obtained using vibrating tube densitometer equipment manufactured by Paar (Paar, Gratz, Austria).

2. Mole fraction (c_M) was calculated from the mass fraction (see Table 2.2 for the conversion formula). (The use of molar quantities means that the mole fractions must be calculated separately at each temperature.)

3. Mean molar volume (V_M/cm^3 mol^{-1}) was calculated from the density data (ρ/kg m^{-3}) using the formula

$$V_M = \frac{1000\left(M_2 c_M + M_1(1-c_M)\right)}{\rho},$$

where M_1 and M_2 are the molecular weights of the solvent and solute, respectively.

4. c_M^2 and c_M^3 were calculated and the spreadsheet regression tool was used to calculate the multiple linear regression between mean molar volume V_M and c_M, c_M^2 and c_M^3. In the case of ethanol in water it was found necessary to do this separately for the water-rich and ethanol-rich parts of the data. The break between the two fits can be seen as a small gap in the curves in Figure 2.20. A single polynomial of high order cannot be used to accurately fit this data because its differential (see later discussion) does not behave well.

5. The predicted V_Ms were compared with the data to check that every data point was accurately predicted by the appropriate polynomial. The slope (dV_M/dc_M) was then calculated by differentiating the polynomial. The method of intercepts entails drawing the tangent to the mean molar volume curve and extrapolating the equation of the tangent to mole fractions of zero to obtain the partial molar volume of solvent and one to obtain the partial molar volume of solute.

$$V_{M_1} = V_M + \frac{dV_M}{dc_M} c_M \tag{2.20}$$

$$V_{M_2} = V_M + \frac{dV_M}{dc_M}(1-c_M). \tag{2.21}$$

2.4.3 APPARENT MOLAR QUANTITIES

In this context, "apparent" means that the quantity to which it is appended is determined at finite concentration and is derived in the form of a finite difference.

2.4.3.1 Apparent Specific Volume

Much of the published literature on ultrasound determination of compressibility of solute molecules employs a definition of "apparent" specific volume (Mitaku et al., 1985; Nomura et al., 1987; Galema and Høiland, 1991), ϕ_{V_2}, which is most succinctly expressed by Sarvazyan (1991):

$$\phi_{V_2} = \frac{V - V_1}{cV}, \tag{2.22}$$

where c is the concentration in g cm^{-3}; V, the total volume of the solution; and V_1, the volume of water to which the solute was added. This equation can be rewritten in terms of density as

$$\phi_{V_2} = \frac{1}{\rho_1} - \frac{\rho - \rho_1}{\rho_1 M}. \tag{2.23}$$

The relationship between density and volume involving molarity is

$$\rho V = MM_2 V + \rho_1 V_1. \tag{2.24}$$

If c is redefined as molarity, then Equation 2.22 can be rewritten, using Equation 2.24 as

$$\phi_{V_2} = \frac{1}{\rho_1} - \frac{\rho - \rho_1}{\rho_1 M}, \tag{2.25}$$

a more convenient form for calculation.

Many authors employ this equation in the form

$$\phi_{V_M} = \phi_{V_m} = \frac{M_2}{\rho} + \frac{1000(\rho_1 - \rho)}{\rho \rho_1 m}. \tag{2.26}$$

This is the "apparent specific molar volume," which can be derived from Equation 2.25 by use of the conversion formulas in Table 2.2.

2.4.3.2 Apparent Compressibility

The "apparent compressibility" as defined by Sarvazyan is

$$\phi_{\kappa_2} = \frac{\kappa V - \kappa_1 V_1}{cV}. \tag{2.27}$$

If c is redefined as molarity then this equation becomes

$$\phi_{\kappa M} = \frac{\kappa V - \kappa_1 V_1}{MV} \tag{2.28}$$

and in terms of density rather than volume it is

$$\phi_{\kappa M} = \frac{\kappa \rho_1 - \kappa_1 \rho}{M \rho_1} + \frac{\kappa_1 M_2}{\rho_1}. \tag{2.29}$$

This equation can be rewritten in terms of the molality m of a solution as

$$\phi_{\kappa M} = \phi_{\kappa m} = \frac{M_2 \kappa}{\rho} + \frac{1000(\kappa \rho_1 - \kappa_1 \rho)}{\rho \rho_1 m}. \tag{2.30}$$

This equation is widely used for the determination of the "apparent molar adiabatic compressibility" of a solute and appears in Harned and Owen (1958).

2.4.3.3 Concentration Increments

Sarvazyan defines three so-called "concentration increments" and shows that a simple identity exists between them:

$$[\kappa] = -2[v] - [\rho], \qquad (2.31)$$

where

$$[\kappa] = \lim_{c \to 0} \frac{1}{\kappa} \frac{\kappa - \kappa_1}{c}; \quad [\rho] = \lim_{c \to 0} \frac{1}{\rho} \frac{\rho - \rho_1}{c}; \quad [v] = \lim_{c \to 0} \frac{1}{v} \frac{v - v_1}{c};$$

c is concentration of the solute, (g cm^{-3}); v is the velocity of compression sound in solution; and v_1 is the velocity of compression sound in the solvent.

The quantities in square brackets are called "concentration increments" by Sarvazyan (1991), and Equation 2.31 provides a means of computing the compressibility from velocity of sound data plotted as a function of concentration.

The advantage of Equation 2.31, according to Sarvazyan, is that it greatly increases the accuracy of the determination of the effects of solvent on solute compressibility. This is because Equation 2.31 depends only on the gradients of sound velocity and density and not on their absolute values. Sarvazyan (1991) measures concentration increments as the difference between the solution property and the solvent property, as a function of concentration determined simultaneously within a single experimental apparatus, although this important fact is not made clear in the review. Sarvazyan states that differential measurements are one to two orders of magnitude more precise than absolute measurements. This formulation is also valuable because the concentration increment of the compressibility can give important information about the state of solute molecules, according to Sarvazyan (1991).

An alternative (and more common) approach is to determine the solvent properties independently. In this case the advantage of a differential formulation as described by Sarvazyan is lost (see §2.4.5). Indeed, a differential formulation of the problem in the case where solution properties are determined independently may be less accurate.

2.4.4 THE DILUTE LIMIT

Apparent quantities (see earlier discussion) are often extrapolated to the limit of zero solute concentration. This is so that the properties of the ideal isolated solute molecule may be approximated. In this case the quantities are sometimes called "limiting" quantities. The extrapolation employed is usually a linear one (see, e.g., Gekko and Yamagami, 1991, and later discussion). But as Garnsey *et al.* (1969) have pointed out, this procedure is fraught with difficulties, not the least of which is that the approach to the dilute limit is far from a linear one in very many cases. In addition, the differential formulations outlined in the earlier section suffer from inaccuracies due to the subtraction of two large quantities. Sarvazyan's method of measuring these quantities simultaneously in a single apparatus is an effective solution to this problem (see Section 2.4.5).

The scale of the problem of achieving the dilute limit may be better understood from the following figures for ethanol in water. At a mole fraction of 0.000705 (equivalent to a volume fraction of ethanol of 0.2%), the velocity change over pure water is 1.8 m s^{-1}. At this concentration there are approximately 230,000 ethanol molecules in a volume of 10^{-20} m^3. This volume has been chosen as it is the volume defined by the decay length of the thermal wave scattered by a solute molecule in water, and by this measure the solution is very far from dilute. At the limit of resolution of ultrasound velocity measurements of, say, 10^{-6} m s^{-1}, using the most sophisticated interferometric techniques, a concentration of ethanol of less than 0.1 molecules of ethanol per 10^{-20} m^3 may be detected. But for most measurements the practical limit is 0.1 m s^{-1} which is equivalent to a concentration change of more than 10,000 ethanol molecules per 10^{-20} m^3.

2.4.4.1 Partial Specific Volume and Partial Specific Adiabatic Compressibility

An approach which avoids some of the difficulties inherent in the apparent volume and apparent compressibility formulations outlined earlier is given by Gekko and co-workers (Gekko and Yamagami, 1991).

$$\overline{\kappa}_2 = -\frac{1}{V_2}\frac{\partial \overline{V}_2}{\partial p} = \frac{\kappa_1}{V_2}\lim_{c\to 0}\left\{\frac{\frac{\kappa}{\kappa_1} - V_1}{c}\right\}$$

$$\overline{V}_1 = \frac{(\rho - c)}{\rho_1}$$

$$\overline{V}_2 = \lim_{c\to 0}\left(\frac{1-\overline{V}_1}{c}\right) = \lim_{c\to 0}\left(\phi_{v_2}\right). \tag{2.32}$$

Here $\overline{\kappa}_2$ is the partial specific adiabatic compressibility of the solute; $\overline{V}_2$, the partial specific volume; and $\overline{V}_1$, the apparent volume fraction of solvent in solution.

2.4.5 SOUND VELOCITY AND CONCENTRATION— THE URICK EQUATION

The Urick equation can be expressed in terms of partial molar volumes. All that is necessary is to form an identity among the volume fraction of the solute, ϕ,

2.4 THE COMPRESSIBILITY OF SOLUTE MOLECULES

its partial volume, V_{M_2}, the mean molar volume of the solution, V_M, and the mole fraction, c_M, as follows:

$$\phi = \frac{V_{M_2} c_M}{V_M}. \qquad (2.33)$$

The Urick equation presumes a quadratic dependence of sound velocity on concentration. This is the case, even in the dilute limit. It has been shown that the Urick equation must be modified to take into account correctly the effects of scattering. In the very long wavelength limit, for the case of a binary mixture we can rewrite the Urick equation (Equation 2.9) in a way that is consistent with the scattering theory formulation adopted in Chapter 4 (Pinfield, 1996):

$$\frac{1}{v^2} = \frac{1}{v_1^2}\left(1 + \alpha\phi + \delta\phi^2\right), \qquad (2.34)$$

where

$$\alpha = \left[\frac{\kappa_{a_2} - \kappa_{a_1}}{\kappa_{a_1}} + \theta + \frac{\rho_2 - \rho_1}{\rho_1}\right],$$

and

$$\delta = \left(\frac{\kappa_{a_2} - \kappa_{a_1}}{\kappa_{a_1}} + \theta\right)\left(\frac{\rho_2 - \rho_1}{\rho_1}\right) + \frac{2(\rho_2 - \rho_1)^2}{3\rho_1^2}$$

$$\theta = (\gamma - 1)\frac{\rho_2 C_{p_2}}{\rho_1 C_{p_1}} R^2,$$

where

$$R = \left[\frac{\dfrac{\beta_2}{\rho_2 C_{p_2}} - \dfrac{\beta_1}{\rho_1 C_{p_1}}}{\dfrac{\beta_1}{\rho_1 C_{p_1}}}\right].$$

Here v is the velocity in solution; v_1, the velocity in the continuous phase; and ϕ, the volume fraction of the solution occupied by solute. α and δ are scattering coefficients defined earlier. Quantities referring to the continuous phase are subscripted $_1$ and quantities referring to the dispersed phase are subscripted $_2$. β is the volume thermal expansivity (K^{-1}); C_p is the specific heat at constant pressure (J kg^{-1} K^{-1}); and γ is the ratio of the specific heats.

A plot of $(v_1^2/v^2) - 1$ against a quadratic polynomial in ϕ should yield a zero intercept and the coefficients α and δ. The compressibility can then be determined from the linear coefficient α, since only κ_2 is unknown. This hypothesis can be tested against published data and yields very interesting results, summarized in Table 2.3 (Pinfield and Povey, 1997).

The quadratic fit is extremely good in all cases and always gives a zero intercept (Pinfield and Povey, 1997). In the case of the dilute ethanol-in-water data no improvement is obtained with a cubic fit and the standard error of the fit to 13 points is 6×10^{-6}. This is much better than can be achieved with a linear fit to the difference between the velocity in solution and in solvent which is used by some workers. This is very encouraging for the use of the modified Urick equation. The compressibility determined from the linear coefficient in Equation 2.34 agrees well with that obtained by the method of Gekko and co-workers when the thermal contribution is small, given the errors inherent in the comparison. However, agreement with the compressibility determined by the method of Sarvazyan (Equation 2.29) is poor, and this may be due to the magnification of errors inherent in this differential formulation. Differential experimental techniques such as those developed by Elias and Eden (1979), Sarvazyan (1991), Eggers and Funck (1973), and Eggers and Kaatze (1996) must be used for accurate results in these cases.

The term $\Delta\rho/\rho_1 \times \Delta\kappa/\kappa_1 + \theta$ should approximately equal the quadratic coefficient δ in Equation 2.34. The reason for the discrepancy probably lies in concentration dependent multiple scattering of the thermal scattering by the dispersed phase which will affect the quadratic term (δ) and this will be discussed further in Chapter 4. It is for this reason that the linear term (α) is used to determine compressibility.

The advantage of the modified Urick equation approach as described in this section is that it gives the adiabatic compressibility of the dispersed phase. The effects of partial molar volume are accounted for accurately by the method of intercepts, without the need to achieve ever higher dilution and without making assumptions as to the form of the approach to the dilute limit. It can be used equally well in dilute and concentrated mixtures, provided the assumptions on which it is based continue to apply. In particular, the Urick equation (Equation 2.34) accounts for thermal scattering through the θ term, which must be calculated from thermodynamic data. This subject will be returned to in Chapter 4.

2.4.6 DETERMINING THE COMPRESSIBILITY OF SOLUTE MOLECULES—A SUMMARY

Great care must be taken when using differential formulations such as Equation 2.26 and Equation 2.29 to calculate compressibility. If the velocity and density cannot be determined simultaneously within a single experimental apparatus operating in a differential manner, then this approach will normally be greatly in error. A better formulation which avoids this pitfall is Equation 2.32.

Hitherto, the Urick equation has not been successfully used for the determination of the compressibility of solute molecules because it has not correctly accounted for scattering. If the Urick equation is rewritten in a form consistent with scattering

2.4 THE COMPRESSIBILITY OF SOLUTE MOLECULES

TABLE 2.3 Comparison of Methods of Determining Solute Compressibility in Selected Binary Mixtures[a]

Ethanol in water T (°C)	c_M	Intercept	Equation 2.34 α	δ	θ	Equation 2.32 κ_2 (Pa^{-1})	Equation 2.29 $\overline{\kappa_2}$ (Pa^{-1})	$\phi_{\kappa M}$ (Pa^{-1})
0	0.05	0.003	−1.68	2.03	0.18	−2.73E-10	−2.67E-10	−1.04E-11
10	0.05	0.000	−1.23	1.08	0.16	−4.13E-11	−5.44E-11	7.10E-13
20	0.05	0.000	−0.99	1.02	0.16	7.66E-11	5.31E-11	6.30E-12
30	0.05	0.000	−0.78	0.90	0.14	1.64E-10	1.44E-10	1.12E-11
Cyclohexanol in n-heptane								
30	0.49	0.000	−0.24	0.109052	0	7.469E-10	9.13E-10	8.68E-11
Dilute ethanol in water								
5	0.008	0.000	−1.20	−0.95	0.17	−2.723E-11	−1.03E-10	2.19E-11
15	0.008	0.000	−0.95	−0.81	0.16	9.35E-11	1.94E-11	2.09E-11

Ethanol in water T (°C)	$\Delta\rho/\rho_1$	$\Delta\kappa/\kappa_1$	$(\Delta\kappa/\kappa)\times(\Delta\rho/\rho_1)$	r	n
0	−0.14	−1.72	0.24	0.9999957	4
10	−0.14	−1.25	0.18	0.9999072	4
20	−0.16	−0.99	0.16	0.9998088	4
30	−0.15	−0.77	0.11	0.9996204	4
Cyclohexanol in n-heptane					
30	0.14	−0.38	−0.05	1	3
Dilute ethanol in water					
5	−0.14	−1.23	0.17	0.9999998	13
15	−0.15	−0.96	0.15	0.9999998	13

[a] Ethanol-in-water data is from D'Arrigo and Paparelli (1987). Cyclohexanol-in-n-heptane data is from Schaafs (1967). Dilute ethanol-in-water data is from Sakurai et al. (1995). κ is defined in Equation 2.8; κ_2 was determined from α in Equation 2.34; $\overline{\kappa_2}$ is defined in Equation 2.32; $\phi_{\kappa M}$ in Equation 2.29. The first three points were taken in the case of the cyclohexanol-in-n-heptane data to obtain a result at the highest dilution consistent with a quadratic fit.

theory, then the adiabatic compressibility may be determined from the parameter α, in a way that accounts for the effects of thermal scattering.

2.4.7 EXPERIMENTAL DATA ON COMPRESSIBILITY AND ITS INTERPRETATION

Sarvazyan (1991) reviews this subject, indicating that the technique can be used to determine the hydration of biological substances and the state of water in the hydration shell, molecular transitions and interactions, thermodynamics and P–V–T state of biological substances in solution, and the conformational dynamics of proteins. The expectations for the ultrasound velocity technique raised by this review have not been disappointed in view of the increasing number of publications appearing in this field, (see Table 2.4).

2.4.7.1 Protein

The apparent molar volume of a protein in water consists of three contributions

$$\phi_{2V} = \phi_c + \phi_{cav} + \Delta\phi_{hyd}, \qquad (2.35)$$

where ϕ_c is the sum of the constitutive or group volumes; ϕ_{cav} is the volume of the cavity in the molecule due to imperfect atomic packing; and $\Delta\phi_{hyd}$ is the volume change due to hydration (Gekko and Yamagami, 1991). ϕ_{cav} comprises both the incompressible cavity formed from atomic close packing and the compressible void space generated on the random close packing of atoms. This volume has been estimated to be 0.02 to 0.05 ml/g, which corresponds to 3–6% of ϕ_{V_2} (Equation 2.23). Thus, the experimentally determined partial molar compressibility of a protein will mainly be due to the contributions of the cavity and hydration. Other work in this area includes that by Gekko and Hasegawa (1989), Sarvazyan (1991), and Sarvazyan and Hemmes (1979).

The partial compressibility measured with ultrasound is an adiabatic quantity. The corresponding isothermal quantity can be calculated from the standard thermodynamic relation (Zemansky, 1957c)

$$\kappa_T = \frac{1}{\rho}\left(\frac{1}{v_I^2} + \frac{\beta^2 T}{C_P}\right), \qquad (2.36)$$

where β is the volume expansivity (K^{-1}); T, the Kelvin or absolute temperature (K); and C_P, the specific heat at constant temperature (J mol^{-1} K^{-1}).

The possibility also exists of determining the thermodynamic parameters for thermal denaturation of proteins, through changes in the partial molar compressibility such as that accompanying the transition to the 'molten globule' intermediate state (Chalikian et al., 1995). This area could become important because of the relative convenience of the ultrasound technique. Preliminary studies indicate that changes in egg protein during thermal denaturation are indeed

TABLE 2.4 Select Bibliography on Ultrasound Measurements of Apparent Molar Adiabatic Compressibility in a Variety of Solutions

Alcohols	Antosiewicz and Shugar (1984); D'Angelo et al. (1994); Kumar (1991); Rajendran (1993); Sakurai et al. (1994, 1995); Yasunaga et al. (1964)
Alkyl halides and other salts	Conway and Verrall (1966); Freyer (1931); Garnsey et al. (1969); Jha and Jha (1986); Kleis and Sanchez (1991); Miecznik (1993); Mikhailov and Shutilov (1965); Millero et al. (1977); Onori (1988, 1990); Owen and Simons (1957); Pillai et al. (1983); Rowe and Chou (1970); Sakurai et al. (1975)
Amino acids and nucleosides	Buckin et al. (1989); Ivnitskii et al. (1987); Millero et al. (1978); Ogawa et al. (1984a,b)
Carbohydrates	Berchiesi et al. (1987); Fedotkin et al. (1980); Galema and Høiland (1991); Juszkiewicz and Antosiewicz (1986); Kaulgud and Dhondge (1988); Maxwell et al. (1984); Nomoto and Kishimoto (1957); Pryor and Roscoe (1954); Shiio (1958); Smith and Winder (1983)
Drug compounds	Iqbal and Verrall (1989)
Hydrocarbons	Rossini et al. (1953); Wang and Nur (1991)
Polymer solutions	Esquivelsirvent et al. (1993); Fenner (1984); Hassun (1988); Majumdar et al. (1980); Mclure et al. (1994); Paladhi and Singh (1990)
Proteins	Bae (1996); Chalikian et al. (1995); Choi et al. (1987); Gekko and Hasegawa (1989); Gekko and Noguchi (1979); Gekko and Yamagami (1991); Kharakoz (1991); Kharakoz and Sarvazyan (1993); Mitaku et al. (1985); Nomura et al. (1987); Pavlovskaya et al. (1992a,b,c); Sarvazyan (1991); Sarvazyan and Hemmes (1979); Tamura et al. (1993)

detectable (Choi et al., 1987; Keatings and Povey, 1996; Bae, 1996) but that the observed change in ultrasound velocity is not associated with thermal denaturation of ovalbumin (Buttner et al., 1996) as might be expected.

3

MULTIPHASE MEDIA

3.1 APPARATUS

Although it is relatively easy to assemble and build ultrasound velocity measurement apparatus, it is not so simple to build an apparatus to the accuracy demanded by most of the measurements which might be made. In particular, the large temperature coefficient of sound velocity requires accurate temperature measurement to be carried out simultaneously with the velocity measurement. It is much more straightforward, and cheaper in the long run, to purchase equipment ready-made than it is to construct it oneself. In this section, the accuracy, precision and other specifications of apparatus sufficient to carry out the measurements described later are laid out (Table 3.1). The measurements described in this chapter used apparatus adhering to the specifications laid out in Table 3.2 and employing the pulse-echo technique (Chapter 2), except where otherwise mentioned. Table 3.3 lists of manufacturers who produce the requisite apparatus, ready-made.

TABLE 3.1 Accuracy and Precision Requirements for the Measurements Described in Chapter 3

Measurand	Accuracy	Precision
Sound velocity	1 m s^{-1}	0.1 m s^{-1}
Temperature	0.1°C	0.01°C

TABLE 3.2 Specifications for Ultrasound Apparatus to Carry Out the Measurements Described in Chapter 3

Frequency	1 to 3 MHz
Transducer aperture	15 to 25 mm
Sound path length	10 to 100 mm

A reference to a manufacturer or apparatus in Table 3.3 does not amount to an endorsement of the product or a claim that the equipment mentioned certainly matches the specification of Table 3.1. Nor is it meant to be an exhaustive list of all ultrasound velocity measurement apparatus.

When using the apparatus, the following precautions must be adhered to.

1. Clean the acoustic cell thoroughly before use. Some contaminants can have a large effect on sound velocity.

2. Degas the sample in the acoustic cell by immersing it in an ultrasonic cleaning bath. If air is present it will manifest itself in the form of bubbles, which rise to the top of the sample. The effectiveness of degassing can be followed by measuring sound velocity during degassing. A tiny amount of air bubbles can have a large effect on sound velocity and signal strength. It is highly likely that air bubbles will prevent any signal transmission and hence prevent any measurement at all. Viscous samples or samples containing surfactant may resist degassing, centrifugation is the only option in these cases.

TABLE 3.3 Manufacturers of Sound Velocity Apparatus

Manufacturer	Application	Name of equipment	Address
Cygnus Ltd.	Multipurpose sound velocity and temperature determinations	UVM	Dorchester, DT1 1PW, England
Cygnus Ltd.	Emulsion and dispersion stability and soft solids studies	Ultrasound Profiler	Dorchester, DT1 1PW, England
Anton Paar	In-line measurement of alcohol content. Sound velocity and density measurement	Density and sound analyzer DSA48	A-8054 GRAZ Postfach 58 Austria
Canongate	Alcohol content determination	Densicheck	Canongate, Edinburgh EH14 2ER, Scotland
Nusonics	Multipurpose concentration analyzer	Sonic Composition Monitor	Nusonics, Tulsa OK 74116-1602, USA

3. Ensure that the sample is well stirred so that the temperature measured in the sample is exactly that in the sound beam. It is very easy for temperature differences of a few tenths of a degree to appear, bear in mind that a tenth of a degree change in temperature is equivalent to a change in velocity of sound in water at 20°C of 0.3 m s^{-1}.

4. Ensure that the sound velocity reading is calibrated regularly by filling the cell with distilled water and using its known velocity of sound at a given temperature.

5. Cover the sample to minimize evaporative losses. This is particularly important in experiments lasting a long time.

3.2 DETERMINING COMPOSITION IN THE ABSENCE OF PHASE CHANGES

Compositional determination by ultrasound velocity measurement depends on a solution of the Urick equation (Equation 2.9) of the form

$$\phi = f(v, T)$$

so that the volume fraction of material may be determined from a velocity and temperature measurement. An example of this form of solution may be seen in Equation 2.11. This may also be done by calibration. The sensitivity of ultrasound velocity to solute volume fraction varies enormously, depending on the solute and solvent. Since it is a quadratic function, it is possible for there to be two values of volume fraction giving the same velocity of sound, at a given temperature (see for example the graph for air/water mixtures, (Figure 2.17). It is also clear that if the concentration of more than one solute is varying then it will be more difficult to interpret results quantitatively. Moreover, in aqueous solutions, the velocity of sound (Figure 2.18) is itself a quadratic function of temperature. This can result in large variations in sensitivity of sound velocity measurements to solute concentration, over even relatively small ranges of temperature. It may therefore be important to choose the measurement temperature carefully, if that is possible. A commonly adopted approach to calibration is simply to establish a network of measurements of velocity, concentration, and temperature which span the required measurement range. This data may be held in a database and the required concentration read off by interpolation onto the database from measurements of temperature and velocity. Alternatively, a calibration function may be determined which fits the data to the requisite accuracy. It is generally unwise to extrapolate very far beyond the range over which the calibration was originally established. However, data may be obtained with a greater range of validity from a careful application of the Urick equation. Following are some examples.

3.2.1 ALCOHOL

The concentration of ethanol in a mixture of ethanol and distilled water can be obtained with considerable accuracy. Studies at Leeds University indicate that the equation

$$\phi = -1.406477 - 0.0044150T + 0.0010010v \\ + 3.149899T^2 + 0.4686299v^2 \\ + 0.0154184vT \quad (3.1)$$

predicts the volume fraction of ethanol in water from a velocity to within a standard error of 0.001, calculated from the regression, over the range 0 to 15% and over the temperature range 9 to 35°C. The data upon which this calibration is based appears in Figure 3.1. Here it should be noted that as the ethanol percentage increases, so the sound velocity becomes progressively less sensitive to the effects of temperature.

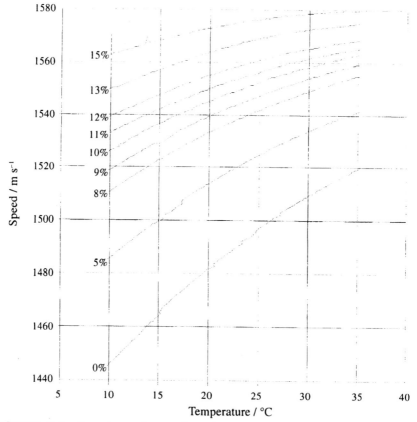

FIGURE 3.1 Dependence of speed of sound on concentration and temperature in ethanol/water mixtures, determined with the Cygnus UVM.

3.2.2 SUGAR

An investigation of the dependence of velocity of sound on concentration in a range of sugars appears in Contreras and co-workers (1992). The density and velocity of sound was measured for a range of sugars and concentrations. A two-way analysis of variance of the velocity and density data produced the following system of equations for velocity and density:

$$v = A_v + B_v x + C_v T + D_v x^2 + E_v T^2 + F_v xT$$
$$r = A_\rho + B_\rho x + C_\rho T + D_\rho x^2 + E_\rho T^2 + F_\rho xT, \quad (3.2)$$

where T is temperature in °C; x, sugar concentration in g/100 ml; v, velocity of sound in m s^{-1}; and ρ, density in kg m^{-3}. The coefficients A–F were determined from fitting the data as shown in Table 3.4.

Equation 3.2 was then used to predict the velocity in mixtures of these sugars, and the total sugar content could be determined to within 0.5% w/v. Since the actual sugar concentration is usually not known in a given fruit juice, an Ultrasonic Sugar Index was employed which assumed that all the sugar was glucose. These studies showed that there was little difference between density, refractive index, and sound velocity methods of determining sugar content. However, the density method was much less sensitive to the type of sugar present. This offers a possible method of determining adulteration by comparison of the sugar-insensitive method (density) with sound velocity measurements.

TABLE 3.4 Regression Coefficients for Glucose, Fructose, and Sucrose in Equation 3.2[a]

	Sugar species		
Coefficient	Glucose	Fructose	Sucrose
A_ρ	1001.2	1001.0	1000.6
B_ρ	3.97	4.06	3.94
C_ρ	−0.116	−0.0754	−0.0359
D_ρ	−0.00403	−0.00267	−0.00115
E_ρ	−0.00217	−0.00383	−0.00438
F_ρ	−0.00419	−0.00751	−0.00507
A_v	1407.0	1405.4	1405.9
B_v	4.41	4.89	3.44
C_v	4.35	4.49	4.46
D_v	0.0151	0.0150	0.0176
E_v	−0.0311	−0.0331	−0.0331
F_v	−0.0371	−0.0495	−0.0319

[a] The coefficients were determined between 10 and 30°C and between 0 and 40 g 100 ml^{-1} sugar concentration.

3.2.3 CONCENTRATION OF A DISPERSED PHASE IN A COLLOIDAL PHASE

A detailed discussion of this subject is contained in the section on ultrasound profiling, Section 3.7 in this chapter.

3.2.4 ANALYSIS OF EDIBLE OILS AND FATS

The general subject of the ultrasonic analysis of edible fats and oils has been reviewed by McClements and Povey (1992). Ultrasound has been applied to the determination of triglyceride composition (Bhattacharya and Deo, 1981; Gouw and Vlugter, 1964, 1966, 1967; Hustad *et al.*, 1970; Javanaud and Rahalkar, 1988; Javanaud *et al.*, 1986; Kuo, 1971; Kuo and Weng, 1975; Rao *et al.*, 1980), dynamic rheology (Gladwell *et al.*, 1985), solid fat content (SFC) of partially crystalline fats (see Section 3.2.5) and the size (see Chapter 4) and concentration (see Section 3.7) of droplets in emulsions.

Javanaud and Rahalkar (1988) suggest a simple empirical formula to relate the ultrasonic velocity of a number of triacylglycerols (triglycerides) to their molecular formula:

$$v = v_0 + av_a + bv_b \tag{3.3}$$

where v_0, v_a, and v_b are constants. v_a corresponds to the increase in velocity per additional carbon atom in the triacylglycerol; v_b, corresponds to the increase in velocity per additional unsaturated bond; a is the total number of carbon atoms in the triacylglycerol molecule; and b is the total number of unsaturated bonds. This equation assumes that the triacylglycerol isomers have similar ultrasound velocities (e.g., POP is equivalent to OPP). Equation 3.3 gives better results than the equation quoted by Gouw and Vlugter (1967; McClements, 1988). McClements and Povey (1988b) suggested that Equation 3.3 should be modified to account for the addition of an unsaturated bond to an already unsaturated fatty acid as follows:

$$v = v_0 + av_a + bv_b + cv_c, \tag{3.4}$$

where c is the total number of unsaturated bonds in the triacylglycerol excluding the first on each unsaturated fatty acid chain and v_c is the velocity increment due to the addition of an unsaturated bond to an unsaturated fatty acid chain. McClements and Povey (1988b) give the data for v_a, v_b, and v_c in Equation 3.4 that is shown in Table 3.5 and Table 3.6.

The variation of ultrasonic velocity with temperature is discussed in Section 3.4.2.1. McClements and Povey (1988b) show that, using Equation 2.15 (the Urick equation for mixtures of components of similar density) and Equation 3.4, accurate predictions may be made of the velocity of sound, simply from a knowledge of the triacylglycerol composition and the data in Tables 3.5, 3.6, and 3.8. In order to illustrate how effective this prediction is, predicted and measured velocities of sound are compared in Table 3.7 for a number of vegetable oils.

3.2 DETERMINING COMPOSITION IN THE ABSENCE OF PHASE CHANGES

TABLE 3.5 Ultrasonic Velocities and Densities of Pure Triacylglycerols at 70°C

Oil	a	b	ρ (kg m^{-3})	v (m s^{-1})
LLL	36	0	887.1[b]	1262.7
PPP	48	0	873.3[b]	1290.2
PSP	50	0	872.3[c]	1292.3
SSS	54	0	870.2[b]	1301.0
POP	50	1	877.6[c]	1293.4
OPP	50	1	877.6[c]	1294.8
POS	52	1	876.5[c]	1297.3
SOS	54	1	875.3[c]	1301.5
OOO	54	3	885.7[b]	1303.5

TABLE 3.6 Velocity Increments in Equation 3.4

Velocity increment	m s^{-1}
v_0	1187.1 ± 3
v_a	2.12 ± 0.07
v_b	0.7 ± 0.4
v_c	3.5

TABLE 3.7 Comparison between Predicted and Measured Ultrasonic Velocities for Vegetable Oils at 70°C

Oil	Measured velocity (m s^{-1})	Calculated velocity (m s^{-1})
Corn	1308.4	1308.2
Grapeseed	1309.3	1308.7
Groundnut	1304.9	1305.9
Olive	1301.5	1302.4
Palm	1298.3	1297.2
Rapeseed	1307.6	1307.6
Safflower	1310.1	1310.4
Soybean	1308.7	1309.3
Sunflower	1310.7	1310.4

3.2.5 CELL SUSPENSIONS

In Chapter 2 it was shown that ultrasound is an effective method for determining the adiabatic compressibility of a dispersed phase in a liquid medium. Self *et al.* (1992) have reviewed ultrasound measurements in fruit and vegetables. The adiabatic compressibility has been determined for suspensions of animal cells, such as red blood cells (Shung *et al.*, 1982) and velocity and attenuation has been determined in suspensions of unicellular plant cells, such as algae and diatoms (Meister and St. Laurent, 1960; Watson and Meister, 1963). Measurements of velocity and attenuation have been made (Self *et al.*, 1992) in suspensions of carrot cells from the taproot cortex. The adiabatic compressibility of these cells was determined to be 2.08×10^{10} Pa^{-1} $\pm 0.03 \times 10^{10}$ Pa^{-1}. The velocity showed little frequency dependence between one and six megahertz. From these data for the single cell and the density of the cell of 1130 kg m^{-3}, Equation 2.8 gives 2061 m s^{-1} for the velocity of sound in a single cell. These data can then be used in more complex models of acoustic propagation in whole plant tissue so that, for example, changes in cell turgor may be measured. The cell suspension method can be used to investigate the effects on cells of changes in their condition. For example, the effects of cell bonding could be studied by the addition of calcium ions to a pectin-containing suspension. Changes in osmotic potential of the continuous phase could be used to alter cell turgor and the effects of changing turgor studied through its impact on compressibility.

3.2.6 TEMPERATURE SCANNING

Measurement of the temperature coefficient of the velocity of sound can be very informative. Examination of Figure 3.7 later, indicates that this will be especially so in systems containing water because of its unusual temperature dependence. For example, at around 70°C, any temperature dependence in the velocity of sound is likely to be due to components other than water, since the temperature coefficient in water is nearly zero at this temperature. At other temperatures, the temperature coefficient is likely to be closely related to the water content. So temperature scanning can be used to determine the water content of mixtures. Data in §2.2.4, §3.2.4, and §3.4.3 are useful in this connection.

3.3 FOLLOWING PHASE TRANSITIONS

3.3.1 GENERAL COMMENTS

Sound speed measurement is an excellent method for following phase transitions in emulsions, dispersions, and homogeneous liquids. The sound speed method can detect not only the temperature at which a transition occurs, but also the proportion of the volume involved in the transition. The basic principles of ultrasound velocity determination of solid/liquid phase transitions are illustrated by reference to a mineral-oil-in-water system. The use of ultrasound velocity

3.3 FOLLOWING PHASE TRANSITIONS

measurement for the determination of phase transitions in systems of industrial interest is explained here by reference to the determination of solid fat content.

Examination of Equation 2.9, the Urick equation, shows that sound speed changes may be expected when either the compressibility or the density or both alter in the material in question. In a solid/liquid or liquid/solid phase transition, often the fractional changes in compressibility are much greater than the fractional changes in density. In Figure 3.2 and later in Figure 3.4 sound speed, density, and compressibility are plotted for a hexadecane oil-in-water emulsion. In these emulsions, the oil phase freezes while the aqueous, continuous phase remains liquid. Results such as these are typical of many mineral oil emulsions. A fuller discussion of these figures appears in the next section. Here attention is drawn to the large change in compressibility on freezing and the first-order transition in

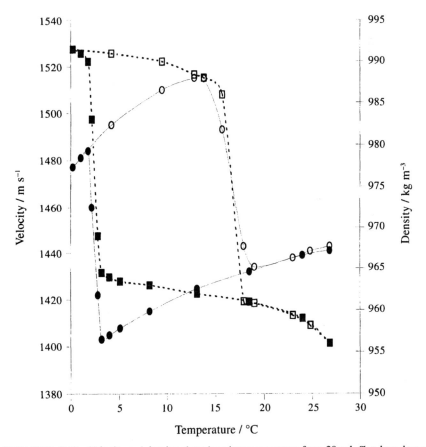

FIGURE 3.2 Velocity and density plotted against temperature for a 20 vol. % n-hexadecane oil-in-water emulsion containing 2 wt % Tween 20. ●, velocity on cooling; ○, velocity on heating; ■, density on cooling; □, density on heating. Average volume-surface particle diameter is 0.8 μm. Data taken from Dickinson et al. (1990).

the sound speed at the point of freezing. This contrasts with the second-order transition in the density at the point of freezing. Consider that the accuracy of sound speed measurement is typically 1 m s^{-1} and the precision 0.1 m s^{-1}, and it is clear from Figure 3.2 that changes in the solid content of the dispersed phase in the emulsion can be detected with an accuracy of better than 1% and a precision of 0.1%.

3.3.2 ATTENUATION CHANGES

It is important to realize, that sound speed can be affected by phase transition phenomena in ways that are far from straightforward. McClements *et al.* (1993b) have shown that when significant amounts of material are participating in a change of phase from liquid to solid and a sound speed measurement is taken, the attenuation changes and associated frequency-dependent velocity changes will be observed. In any material which is partly crystalline and in which the solid and liquid phases are in equilibrium (*solid* ⇔ *liquid*), a small amount of energy may be sufficient to disturb this equilibrium. From Table 2.1 it can be seen that associated with the ultrasonic pressure wave is a thermally induced pressure wave whose amplitude is equivalent to a few millikelvins. These fluctuations in temperature are sufficient to perturb the equilibrium, so that the proportion of solid to liquid phase can oscillate around the equilibrium value, over a range of frequencies of the order of (relaxation time)$^{-1}$ (Bulanov, 1979). This process transfers energy from the pressure wave into fluctuations in the solid/liquid system, increasing attenuation and causing frequency-dependent changes in ultrasonic velocity. This process will occur so long as the relaxation time for the *solid* ⇔ *liquid* fluctuation is shorter than, or of the order of, the period of the ultrasound wave. Thus, it may be possible to quantify the relaxation time of the *solid* ⇔ *liquid* process and to measure the quantity of material participating at any moment in the phase transition. This requires a measurement of attenuation (Figure 3.3) and sound speed (Figure 3.2).

The interested reader is referred to Akulichev and Bulanov (1982), Bulanov (1979), Dobromyslov and Koshkin (1970), Glazov and Kim (1988), Gorbunov *et al.* (1966), McClements *et al.* (1993b), Perrin (1981), Raman (1982), and Tiddy *et al.* (1982) for a full treatment of this subject. Very interesting work on ice has also been published (Hiki and Tamura, 1981; Tamura *et al.*, 1986) which also, unfortunately, is beyond the scope of this work.

For the purposes of this discussion, the important point to note is that the Urick equation *may not* apply when there is the possibility of a *solid* ⇔ *liquid* transition. This is a particularly important point with regard to fat crystallization. *It is necessary to ensure that the phase transition process is complete* before attempting a quantitative analysis of the amount of solid material on the basis of sound speed data. In practice this amounts to demanding that the measurement be made isothermally. However, it may be presumed, although it has yet to be definitely proved, that in the case of phase transitions where the solid and liquid

3.3 FOLLOWING PHASE TRANSITIONS

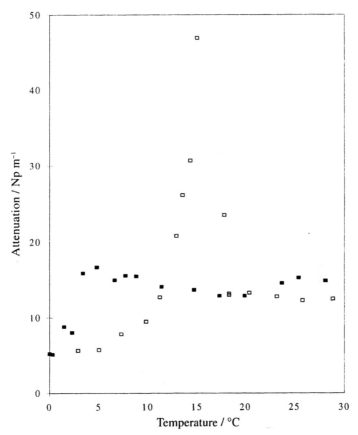

FIGURE 3.3 Attenuation changes in a hexadecane oil-in-water emulsion during freezing and melting of the oil phase. ■, cooling; ☐, heating.

phases are energetically very close and the relaxation frequency between them is higher than the ultrasound frequency, the ultrasound will interact strongly with the phase transition in any case.

A second important caveat is that *any* phase transition is a potential source of attenuation and dispersion. This makes ultrasound a valuable and much underused tool in the study of phase transitions. For instance, ultrasound measurements can make a fundamental contribution to the study of protein conformation (see Chapter 2).

3.3.3 CRYSTALLIZING SOLIDS

The direct determination of solids content from ultrasound velocity, in the absence of thermal relaxation phenomena, is based on the Urick equation (Equation 2.9). As an example of the use of these equations, the determination of the solids content of an oil-in-water emulsion is shown later in Figure 3.5, together with the corresponding velocity data (Dickinson et al., 1990).

First of all, a regression line is calculated for that part of the data which applies to the sample in a wholly liquid state. This is the liquid line in Figure 3.5. In this case the liquid-only data was fitted using multiple linear regression with the following equation:

$$v = 1394.541 + 2.792138T + -0.03954T^2. \qquad (3.5)$$

The entire calculation was carried out using a standard spreadsheet package; the regression coefficient was 0.99978, equivalent to a standard error of 0.38 m s^{-1} on eight observations.

The same procedure was carried out with respect to the data from the sample in the completely solid state, giving a regression equation of the form

$$v = 1475.958 + 5.258264T + -0.17426T^2. \qquad (3.6)$$

In this case the multiple regression was 0.99929 and the standard error was 0.62 m s^{-1} on four observations. The reduced number of samples reflects the reduced range over which the sample was fully solid, as compared with the wholly liquid material. A selection of the experimental data has been added to the graph in order to relate the calculated curves to the measured data. The solid content is then calculated using

$$\phi = \frac{\frac{1}{v^2} - \frac{1}{v_l^2}}{\frac{1}{v_s^2} - \frac{1}{v_l^2}} \Phi_s. \qquad (3.7)$$

Here ϕ is the calculated *volume fraction* of solid present in the entire system; Φ_s, the volume fraction of the dispersed phase; v the measured velocity of sound; v_s, the extrapolated velocity of sound for an emulsion containing wholly solid droplets, and v_l, the extrapolated velocity of sound for an emulsion comprising wholly liquid droplets.

Equation 3.7 was first derived by Miles et al. (1985) using Taylor series expansions of the compressibility and density in the Wood equation (Equation 2.8). McClements (1988) showed that the same equation could be derived directly from the Urick equation (Equation 2.9) if it was assumed that fractional difference in density between phases were much smaller than fractional differences in compressibility. This is in fact generally the case; in Figure 3.4 (which was derived from the data in Figure 3.2 and Figure 3.5) it can be seen that the compressibility changes are an order of magnitude greater than density changes in the case of hexadecane.

Pinfield et al. (1995) showed that the Urick equation was a special case of scattering theory and demonstrated that Equation 3.7 could be derived directly from scattering theory. When derived in this way, it becomes apparent that all the

3.3 FOLLOWING PHASE TRANSITIONS

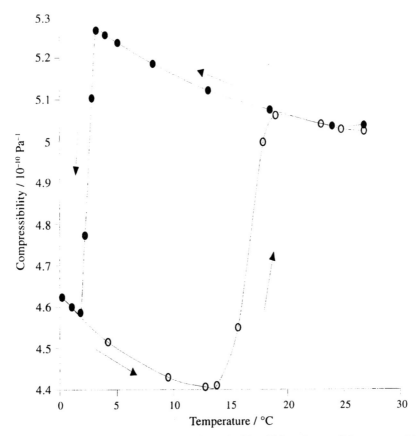

FIGURE 3.4 Changes in the compressibility of a 20 vol% hexadecane oil-in-water emulsion undergoing freezing and melting. Droplet diameter was 0.8 μm. (From Dickinson *et al.*, 1990). ●, cooling; ○, heating. The compressibility plotted here is determined from the velocity and density plotted in Figure 3.2 using Equation 2.8.

first-order scattering terms cancel out and that this equation is relatively insensitive to scattering.

In this section the fundamentals of determining solids content in crystallizing systems have been addressed. Not surprisingly, the actual practice in many situations can be a good deal more complicated. These situations will be considered in the next section.

3.3.4 CRYSTALLIZATION IN COLLOIDAL SYSTEMS

Ultrasound velocity measurement is a new and successful method for following crystallization in colloidal systems. In the last few years this technique has been developed to a high degree (McClements *et al.*, 1990a, 1993a; Coupland *et al.*,

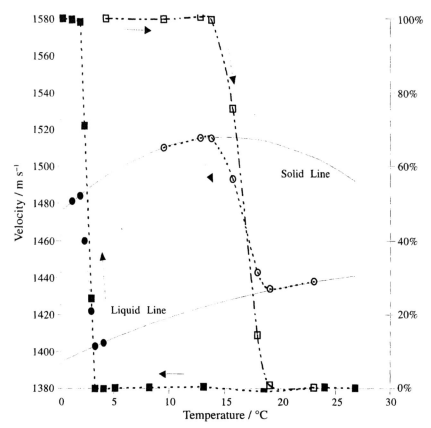

FIGURE 3.5 Velocity of sound and solids content for the data shown in Figure 3.4 (from Dickinson *et al.*, 1990). ●, selected sound velocity data for cooling; ○, selected sound velocity data for heating; ■, solids content for cooling; □, solids content for heating.

1993; Povey, 1995; McClements and Povey, 1992; Dickinson *et al.*, 1991, 1996). The technique has been used to investigate isothermal inhomogeneous surface nucleation in mixtures of solid and liquid mineral droplets (McClements *et al.*, 1993a; Dickinson *et al.*, 1993b, 1996). It is particularly effective for determining nucleation kinetics in systems with a sharp melting point.

3.4 DETERMINATION OF SOLID FAT CONTENT

3.4.1 INTRODUCTION

There is a need for a simple, cheap, and accurate means of measuring solid fat content in fat-containing emulsions containing low levels of solid fat (Wokke and van den Wal, 1991). This can be done by NMR measurements, but the equipment

3.4 DETERMINATION OF SOLID FAT CONTENT

required is expensive and sensitive to disturbance. In the spreads industry, solid fat content is expressed as the so-called "N-value" for the weight percentage of solid fats at a specified temperature. Determinations were originally performed by the so-called "dilatation" method in which differences of volume in solid and liquid fat enabled the solids content to be calculated. Over the past 20 years it has become fairly common to use an NMR technique, and thereby obtain N-values directly from a sample.

The ultrasound velocity technique gives the volume fraction of solid contained in the whole system under study. In a comparison of pulsed nuclear magnetic resonance (pNMR) and ultrasonic velocity techniques for determining solid fat contents, McClements and Povey (1988a) showed that the precision of the ultrasonic technique (0.2%) and the pNMR weight method (0.3%) were significantly better than that of the pNMR direct method (0.7%). More recent, unpublished work in the author's laboratory has shown that the ultrasonic technique can be over 20 times more sensitive than pNMR for 20% oil by volume, hardened palm oil-in-water emulsions (see §3.4.7). Add to this the facility with which the technique can be translated into in-line measurement within a factory, and the result is a powerful technique for monitoring solid fat content. The technique has been studied for a number of years (Hussin, 1982; Povey, 1984; Hussin and Povey, 1984; McClements and Povey, 1987, 1988a; Miles *et al.* 1985).

3.4.2 GENERAL METHOD

The essentials of determining solid fat content (SFC) by ultrasound velocity measurement are the same as those outlined earlier for determining the amount of crystallized material. In Figure 3.6, the variation of ultrasound velocity for a mixed triacylglycerol (triglyceride) oil-in-water margarine is shown in which the solid content of the oil phase varies from 0 to 100%. Three regions can be identified (McClements and Povey, 1988a). In region I the triacylglycerol is completely solid and the decrease of velocity with increasing temperature is due to the dominant effect of the solid fat. As the temperature rises further (region II), the triacylglycerols melt and the velocity decreases more rapidly than in region I. Once the triacylglycerol reaches a temperature where it is completely liquid (region III), the rate of decrease in velocity with temperature reduces again. Exceptions to this general picture can arise due to the solubility of one solid triacylglycerol in another, for example, POS and POP in paraffin.

Varying triacylglycerol composition may affect the solids content determined by ultrasound velocity. The effects of varying oil composition have been considered in Section 3.2.4.

3.4.2.1 Region I

The variation of velocity with temperature can be described using Taylor series expansions (Miles *et al.*, 1985). A better fit may be obtained in the case of mixtures of triacylglycerols using an exponential function:

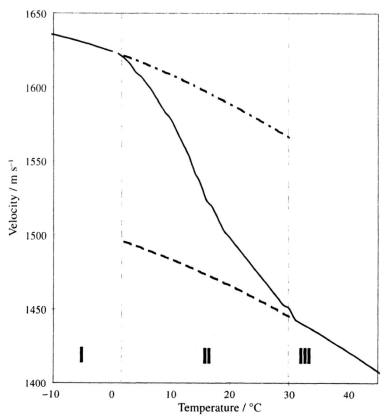

FIGURE 3.6 Variation of ultrasound velocity for a mixed triacylglycerol oil-in-water margarine. The emulsion includes 80% oil. In region I the triacylglycerol is completely solid, in region III it is completely liquid, and in region II it is part solid and part liquid. (Adapted from McClements and Povey, 1988a.)

$$v = \alpha_T \exp(\beta_T T), \tag{3.8}$$

where v is the velocity (m s^{-1}), T is the temperature (°C), and α_T and β_T are experimentally determined constants. Values for α and β for a range of oils appear in Table 3.8 and those for pure triacylglycerols dispersed in paraffin are listed in Table 3.9.

An important point is that the type of solid triacylglycerol present does not greatly influence the velocity through the sample; all the mixtures in Table 3.9 have a variation of velocity with temperature of the order of -3.8 m s^{-1}. This is not the case with distearin (SS) and stearic acid (SA).

3.4.2.2 Region III

Equation 3.8 may be used for the wholly liquid oil as well as for the wholly solid oil, and the corresponding values of a and b appear in Table 3.9. Although

3.4 DETERMINATION OF SOLID FAT CONTENT

TABLE 3.8 Variation of Ultrasonic Velocity with Temperature for Nine Vegetable Oils[a]

	T (°C)	n	α_T m s^{-1} ± 0.2	β_T °C^{-1} ± 0.00002	r
Corn	20–70	36	1539.3	−0.00232	−0.9999
Grapeseed	20–70	36	1540.9	−0.00233	−0.9999
Groundnut	20–70	36	1535.8	−0.0233	−0.9995
Olive	20–70	36	1536.1	−0.00237	−0.9996
Palm	50–70	10	1529.2	−0.00234	−1.0000
Rapeseed	20–70	36	1539.8	−0.00234	−0.9999
Safflower	20–70	36	1541.3	−0.00232	−0.9999
Soybean	5–70	30	1539.6	−0.00232	−1.0000
Sunflower	5–70	30	1541.5	−0.00232	−1.0000

[a]McClements and Povey (1988a).

the temperature coefficient of velocity is similar (~−3.3 m s^{-1}) in all cases, the absolute velocity varies significantly. However, the Urick equation may be used to predict the velocity in an oil of known composition (§3.2.4).

3.4.2.3 Region II

The melting of the solid triacylglycerol results in a decrease in velocity additional to that due to the temperature coefficients of regions I and III. The varying solubility of acylglycerols in paraffin oil affects the velocity in Region II. The solubility can be described by the Hildebrandt equation for ideal solutions:

$$\ln(x) = \frac{\Delta H_f}{R\left(\dfrac{1}{T} - \dfrac{1}{T_{mp}}\right)}, \qquad (3.9)$$

where x is the mole fraction of acylglycerol in the liquid phase, ΔH_f is the enthalpy change per mole of crystallizing material, R is the gas constant, T_{mp} is the absolute melting point of the pure acylglycerol, and T is the absolute temperature (K). This equation applies to high-melting-point fats dissolved in low-melting-point oils. The mass fraction (w) of acylglycerol in the solid state at any temperature in region II may then be calculated from

$$w = \frac{w_t - \dfrac{x(1-w_t)M_t}{(1-x)M_s}}{w_t}, \qquad (3.10)$$

where w_t is the total mass fraction of acylglycerol present in the mixture; M_t is the molecular weight of the acylglycerol phase; and M_s is the molecular weight of the solvent. The total solid content of acylglycerols can be determined independently

TABLE 3.9 Variation of Ultrasound Velocity with Temperature for Regions I and III for 15% w/w Triacylglycerol/Paraffin Mixtures[a]

Region I

Glyceride[b]	n	α_T (± 0.5 m s^{-1})	β_T (± 0.00007°C^{-1})	r	dv/dT (± 0.1 m s^{-1}°C^{-1})
SSS	7	1594.5	−0.00247	−0.999	−3.8
PPP	7	1595.4	−0.00246	−0.997	−3.8
LLL	4	1593.4	−0.00247	−0.999	−3.9
PSP	6	1594.2	−0.00243	−0.999	−3.7
POP	—	—	—	—	—
SOS	—	—	—	—	—
POS	—	—	—	—	—
SS	11	1599.1	−0.00252	−0.995	−3.8
SA	11	1614.5	−0.00260	−0.993	−4.0
Paraffin Oil					

Region III

Glyceride[b]	n	α_T (± 0.5 m s^{-1})	β_T (± 0.00007°C^{-1})	r	dv/dT (± 0.1 m s^{-1}°C^{-1})
SSS	4	1545.5	−0.00248	−0.999	−3.3
PPP	4	1545.3	−0.00249	−0.997	−3.3
LLL	8	1534.3	−0.00244	−0.999	−3.3
PSP	—	—	—	—	—
POP	6	1537.9	−0.00242	−0.999	−3.3
SOS	6	1539.0	−0.00241	−0.999	−3.2
POS	6	1540.6	−0.00247	−0.999	−3.2
SS	5	1553.6	−0.00254	−0.995	−3.3
SA	10	1540.7	−0.00249	−0.998	−3.3
Paraffin Oil	15	1546.6	−0.00247	1.000	−3.5

[a]McClements and Povey (1988b).
[b]SSS, tristearin; PPP, tripalmitin; LLL, trilaurin; PSP, glycerol 1,3 dipalmitate-2-stearate; POP, glycerol 1,3 dipalmitate-2-oleate; SOS, glycerol 1,3 distearate-2-oleate; POS, glycerol 1-palmitate 2-oleate-3-stearate; SS distearin; SA, stearic acid.

from Equation 3.7 and McClements and Povey (1988a) show that good agreement is obtained between the ideal solubility equation (Equation 3.9) and the velocity of sound determination of solid content. The variation in density can be ignored.

3.4.3 MARGARINE

One practical point associated with the determination of solid fat content is that it is often difficult to determine the velocity of sound in the solid phase. There are two reasons for this. First of all, it is difficult to measure the velocity of sound in pure solid fats. This is because it is difficult to maneuver them into a geometry where the acoustical path can be measured accurately. It is also difficult to get accurate readings because of the tendency for voids to form within the fat as it solidifies. An account of the measurement of ultrasound velocity in solid fat can be found in McClements (1988). It is preferable to determine the velocity of sound in the solid fat, dispersed in a liquid such as water or mineral oil, in which it is insoluble. However, dispersion of the fat in water considerably lowers the temperature at which the fat solidifies, very frequently to such low temperatures that measurement becomes impractical. McClements (1988) and McClements and Povey (1988b) have shown that a wide range of fats share common features; these make it possible to predict the velocity of a solid fat, without actually measuring it. For fats with melting points below about $-15°C$, it is usually necessary to fit the solid line using a combination of methods.

This is most conveniently done with a spreadsheet program and an example of a sunflower-oil-in-water margarine is shown in the Figure 3.7. These data are taken from Kelly *et al.* (1990). The oil is a blend of sunflower oil with hardened palm oil and other vegetable oils, which are added to give a sharp melting point for the fat when it enters the mouth. In this type of margarine, the bulk of the fat remains liquid at all processing and storage temperatures. The solids determined by ultrasound agree well with those determined from pNMR. Note that the addition of salt, whey solids, and other ingredients to the aqueous phase has a considerable effect. For this reason it is generally desirable to measure the velocity of sound in the aqueous phase only. This can be done, before it is added to the premix, or it can be extracted postprocessing by centrifugation or possibly filtration. The liquid line is obtained for the same margarine as in §3.3.3, by fitting a polynomial to that part of the velocity data for which the sample is wholly liquid. The solid line is obtained by assuming that it fits the equation given in Table 3.10 (McClements, 1988; Kelly *et al.*, 1990). An alternative approach would be to use the data of Table 3.9 and the method outlined in §3.1.

3.4.4 CHOCOLATE

Measuring solid fat content in chocolate and cocoa butter is complicated by the hardness of the fats, which makes it impossible to make measurements in regions I, II, and III with a single apparatus. To circumvent this problem, hard fats need to be dispersed in a liquid medium, generally a mineral oil such as paraffin, in which the fat is insoluble. One method of dispersion is to melt the hard fat and then mix it with the liquid oil. This has the disadvantage that the fat must then be taken through its full tempering cycle in order to obtain its SFC. A similar method

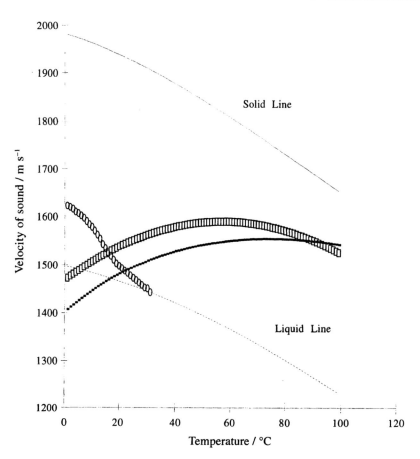

FIGURE 3.7 Velocity of sound in sunflower oil-in-water margarine and its components, plotted against temperature. O, Margarine premix — 80% fat; - - -, pure water; □, aqueous phase.

is to disperse the fat as an oil-in-water emulsion; this was the procedure followed in the case of the hardened palm oil data referred to in §3.4.2 and §3.5.1. An alternative method developed in the laboratories at Leeds consists of dispersing the fat as a powder in mineral oil. The fat is cooled to 0°C in order to harden it,

TABLE 3.10 Physical Properties of Margarine Component Phases

	Velocity (m s^{-1})	Density (kg m^{-3})
Aqueous phase	$1531.7 + 2.93(T - 17.6) - 0.0368(T - 17.6)^2$	1041
Liquid oil phase	$1378.2 - 3.14(T - 44.6) - 0.0066(T - 44.6)^2$	920
Solid fat phase	$2190 \exp(-0.00264T)$	1020

and then it is powdered using a file. The powdered particles must be smaller than the wavelength of sound (0.7 mm at 2 MHz in Figure 3.8) for this procedure to work. The particles are kept dispersed at a volume fraction of 20% w/w by stirring with a magnetic flea and a submersible stirrer. Higher-volume fractions become solid at lower temperatures, making accurate velocity measurement impossible. The sample is then warmed to the measurement temperature and held there until the velocity of sound becomes constant. This is taken as an indication that the solid fat content has ceased to change. The effect of nonfat solids, such as sugar and cocoa solids, is removed by making the measurements as a function of temperature and following the transformation of the fat to oil and/or vice versa. The results of such experiments with eating chocolate are shown in Figure 3.8.

One very puzzling feature of these data is that it was necessary to raise the temperature of the sample above 80°C before liquid-like behavior became apparent. This can be explained by the presence of high-melting-point phospholipids in the cocoa butter. However, although these phospholipids are present in variable quantities in cocoa butter, they are not generally present at concentrations above about 0.13% (Gunstone et al., 1994). It may be that phospholipids have a disproportionate effect on the sound velocity.

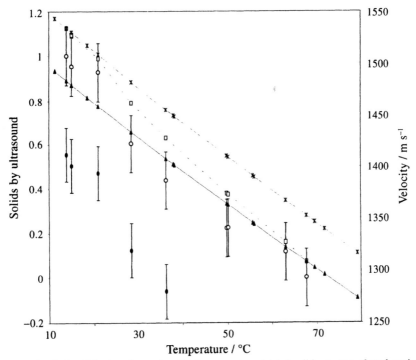

FIGURE 3.8 Velocity of sound, solid fat content, and total solids content plotted against temperature for eating chocolate. -x-, solid line; -▲-, liquid line; ...□..., velocity of sound; ○, total solids; ●, solid fat content.

The liquid line shown in Figure 3.8 was obtained from data above 82°C. The solid line was obtained using the data of Table 3.9 and assuming that the POS data most closely fitted the behavior of cocoa butter. The total solids was then calculated from the liquid and the solid line using Equation 3.7, as before. The solid fat content was then obtained by subtracting the solid fat content at 32°C from the total solids obtained at lower temperatures, the assumption being made that the solid fat content of the chocolate/cocoa butter was zero above 32°C.

The continuous change in total solids up to 68°C suggests that if phospholipids are indeed responsible, then they are continuously melting over the entire temperature range. The solid fat content agrees reasonably well with pNMR measurements of chocolate.

3.4.5 ACCURACY

As the proportion of the sample occupied by oil falls, so the overall accuracy of the solid fat determination will fall. In Figure 3.9, the solids content determined

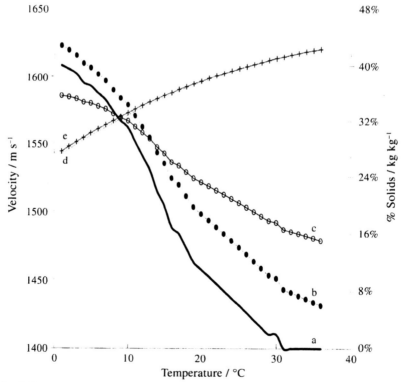

FIGURE 3.9 Solids content as a function of temperature: (a) Solids as a proportion of the total (kg/kg); (b) velocity in emulsion for 80% w/w oil; (c) velocity in emulsion for 60% w/w oil; (d) velocity in emulsion for 10% w/w oil; (e) velocity in 10% emulsion in the absence of crystallization.

3.4 DETERMINATION OF SOLID FAT CONTENT

for an 80% oil-in-water emulsion is used to predict the velocity in emulsions containing smaller amounts of oil, and to estimate the accuracy. At 80% oil, the accuracy of the SFC determination was estimated to be better than 0.1%. Examination of the curve for the 10% data indicates that the difference between the data for material containing 40% solid in the oil phase and the same emulsion at the same temperature in the absence of solid is very small. Yet even here the accuracy is better than 2%. At 40% solids in the oil, there is 2.2 m s^{-1} between the liquid line and the emulsion containing solid and the technique operates with a precision of 0.1 m s^{-1}. The total oil occupies only 10% of the sample, so the solid occupies 4% and a change in that solid content of 2% is easily detectable.

3.4.6 ANOMALIES CLOSE TO THE MELTING POINT

In Figure 3.10 are shown data for an emulsion comprising hardened palm oil, emulsified in water using Tween20 surfactant and containing 10% oil in the dispersed phase.

A sharp melting transition associated with hardened palm oil can be seen between 53°C and 58°C. On the cooling cycle, the supercooling due to oil

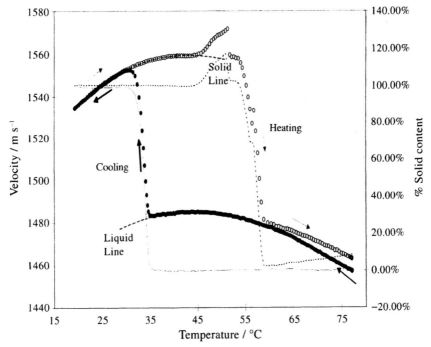

FIGURE 3.10 Ultrasound velocity and solids content in a 10% hardened palm oil-in-water emulsion, 2% Tween 20 surfactant. ●, velocity as emulsion cools; ○, velocity as emulsion heats; —, solids content calculated from velocity cooling cycle data; - - -, solids content calculated from velocity heating cycle data (Kloek, 1995).

dispersion is clear and is similar to that observed in the mineral oil dispersion of Figure 3.4. An as-yet-unexplained effect (but see §3.3.2) is seen on the heating cycle, where the solid content appears to rise above 100%, just prior to melting. This may be due to a phase transition in the palm oil. The emulsion was not stable on the heating cycle because of due to particle coalescence, and the rising solids content indicated above the palm oil melting point is an artifact, caused by changes in the particle size distribution. The precision in this case is amazingly high. With the sample containing only 10% oil, we can observe changes in the solid content in that oil phase to better than five parts in one thousand of oil.

3.4.7 COMPARISON WITH DILATOMETRY AND PULSED NUCLEAR MAGNETIC RESONANCE

The major contemporary methods of measuring solid fat content are dilatometry (A.O.C.S., 1973), calorimetric (Walker and Bosin, 1971), wideline nuclear magnetic resonance (Mansfield, 1971) and pulsed nuclear magnetic resonance (pNMR) (Waddington, 1986). None of these methods are easily adapted from laboratory conditions for use under hostile industrial conditions.

The advantages of ultrasound velocity measurement for solid fat content determination are as follows:

- It is easily adapted to in-line measurement in the factory, as well as the laboratory bench
- It can give readings as rapidly as 10 per second
- It is relatively low cost
- It can give reliable and very accurate results
- The transducer is small and robust, and cheap to replace if damaged, and the associated instrumentation is simple

Although NMR methods have given excellent results at relatively high solid fat levels (above 10% solids), they have three major disadvantages. First, NMR is difficult to apply in-line; NMR apparatus is bulky and sensitive to interference from a number of sources. Second, on-line NMR apparatus is generally prohibitively expensive. Third, NMR measurements show wide margins of error at low solid-fat levels. NMR measurements become increasingly inaccurate as solid fat levels fall below about 4 wt% of the emulsion system. This is the very region in which rapid and trustworthy solid fat measurements are required for accurate control of the spreads manufacturing process. However, pNMR does have the advantage of ultrasound velocity measurement that it is insensitive to the polymorphic form of the fat.

A direct comparison of pulsed NMR and the ultrasound velocity technique is shown in Figure 3.11. The ultrasound velocity changes by approximately 70 m s^{-1} on crystallization, the corresponding change in pNMR signal is 6.5. The precision of the ultrasound velocity is 0.1 m s^{-1}, and that of the pNMR signal is 0.2. So the

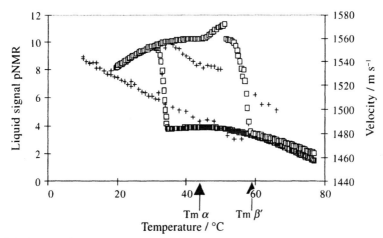

FIGURE 3.11 Cooling–heating cycle of 20% hardened palm oil-in-water emulsion measured by the liquid pNMR signal (+) and the ultrasound velocity (□). d_{32} = 0.54 µm, emulsifier was 2% w/w sodium caseinate (Kloek, 1995).

ultrasound velocity determines the solid content in the dispersed oil phase of this emulsified system to an accuracy of about 0.15% whereas the pNMR is accurate to within about 3.5%.

Although the ultrasound velocity technique is generally the best method for monitoring solid fat content, it is not necessarily always the best method. The ultrasound technique works best for the assessment of fats with a narrow melting range. Under laboratory conditions, it is better adapted than pNMR for the assessment of emulsified fats because the water component can be determined independently, from the temperature dependence of the velocity of sound away from the fat melting point. Ultrasound is better at very low and very high solid contents, whereas pNMR works best at intermediate fat contents. By measuring the relaxation times of the proton in a solid and a liquid environment, pNMR measures a fundamentally different quantity than the compressibility, which underlies the ultrasound determination of solids. Thus, if other solids, not fat, are also undergoing a solid–liquid transition, they may have a disproportionate effect on the ultrasound velocity, giving a misleading figure for the solids content. Also, as pointed out in the previous section, melting material can alter the sound velocity through a relaxation process. Thus it is possible to obtain anomalous results from an ultrasound solid fat determination in some circumstances. This may particularly be the case in materials which have been blended to give a wide range of melting points. In this case, the overall sensitivity of the technique is reduced at any one temperature to the proportion of the total material that is melting at that temperature. Under these circumstances relaxation processes may introduce large errors, and it is necessary to take great care to stabilize the sample isothermally before taking a measurement. It is useful to monitor attenuation, which rises when relaxation processes contribute to the velocity of sound.

Finally, although the pNMR technique is insensitive to the polymorphic form of the fat, the ultrasound velocity technique is affected by polymorphism, although the degree to which this is so has still to be determined.

In practice both the ultrasound and pNMR techniques are superior to the dilatometry techniques which are still most widely used. Generally, dilatometry is a very labour-intensive and error-prone operation. However, both pNMR and ultrasound (especially ultrasound) are relative newcomers to the industry and have still to win wide acceptance outside the larger industrial laboratories and concerns.

3.4.8 SOLID CONTENT AND PARTICLE SIZE

Particle size, in general, has a very significant effect on ultrasound velocity. If sound velocity is sensitive to both particle size and concentration, how are we able to determine concentration without a knowledge of particle size? The procedure of measuring the velocity in the emulsion, first when it is fully liquid and then finally when solid appears in the system, works well so long as the particle size distribution does not change. Accurate solid content values can only be obtained in an emulsion which is scattering sound if the procedure laid out in §3.3.3 is closely followed. This method accurately cancels out the effects of particle size under most circumstances. In addition, the effects of particle shape are also probably very small, because the scattering in most emulsion systems is dominated by thermal effects which are primarily monopolar (Chapter 4).

An example of the effects of changing particle size may be seen in Figure 3.11, in which the slightly raised velocity on the final part of the heating cycle is due to droplet coalescence. The coalescence became apparent towards the end of the experiment in the form of an oil layer visible on the top of the sample. Particle size effects are often seen in experiments with the ultrasound profiler, where fractionation of particles on the basis of their size may be particularly pronounced in the cream layer.

3.5 CRYSTAL NUCLEATION

3.5.1 CRYSTAL NUCLEATION RATES

The high sensitivity of the ultrasound velocity method to the presence of crystallizing solids makes it an ideal technique for the study of crystal nucleation rates. The method consists of dispersing the crystallizable material in liquid form as an emulsion. The emulsion is then cooled until crystallization begins, and then the rate of appearance of solid material is measured as a function of temperature around the crystallization temperature. An example of such a measurement is given in Figure 3.12.

In the experiment of Figure 3.12 it took 4 to 5 min for the sample to reach thermal equilibrium. During the bulk of this time no crystallization will occur

3.5 CRYSTAL NUCLEATION

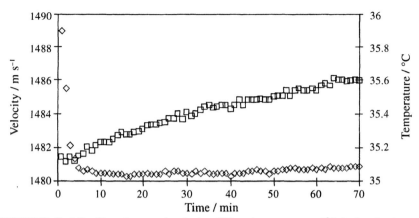

FIGURE 3.12 The ultrasound velocity (□) and temperature (◇) during isothermal crystallization of 20% hardened palm oil-in-water at 35.1°C (Kloek, 1995).

because the temperature is too high. However, for the highly supercooled samples some crystallization will occur during cooling, which will create an error at the beginning of the crystallization process, although not at the end. After a series of such measurements over a range of temperature, a series of plots such as those in Figure 3.13 are obtained. The quality of the data is good enough for different models of nucleation to be tested.

Comparison of Figure 3.13 with Figure 3.14 indicates that the heterogeneous nucleation model gives a much better fit at longer times, but that the initial stages of crystallization appear to be due to homogeneous nucleation. Nucleation rates

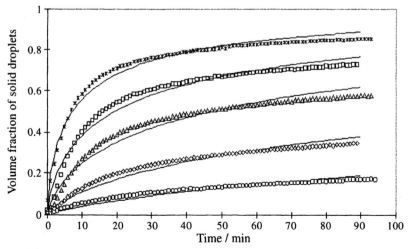

FIGURE 3.13 Isothermal crystallization curve for 20% hardened palm oil-in-water emulsion. The continuous lines are fits to a homogeneous crystal nucleation model.

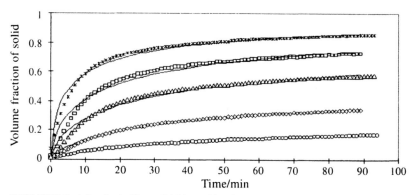

FIGURE 3.14 As in Figure 3.13 but with a heterogeneous nucleation model fitted.

can be calculated from data such as these and detailed comparisons of various models of nucleation made, including the effects of particle size distribution.

This is not the place for a discussion of crystal nucleation theory. However, the quality of the data available from the ultrasound velocity technique is an extremely valuable addition to the study of crystallization.

3.5.2 ICE

Although it is difficult to measure the transformation of bulk water into ice and vice versa, the techniques developed for studying fats can be used to study ice nucleation. In this case water must first be dispersed as an emulsion in a liquid oil phase. The nucleation of ice can then be studied in the same way as the nucleation kinetics of hardened palm oil (§3.5.1). Such measurements have been made (Archer *et al.*, 1996). An emulsion of double-distilled water-in-oil was prepared with a volume fraction of water of 0.2 and a continuous oil phase comprising 5% w/w lanolin in light white mineral oil. The resultant emulsion had a mean droplet diameter of 8 μm, measured using optical microscopy. Bacterial suspensions comprising 0.001% w/w SNOMAX, a commercially available freeze-dried preparation of ice nucleation active *Pseudomonas syringae* 31 A bacteria, were added to one emulsion sample and pure double distilled water formed the dispersed phase of the second emulsion. A third emulsion had another ice nucleator, silver iodide (AgI), added to the dispersed phase.

Measurements were made in a UVM ultrasound cell which was placed in an ethylene glycol/water bath with associated cooling and heating devices (Grant, U.K.). The cooling and warming rates were preprogrammed using a temperature programmer model PZ1 (Grant, U.K.). As usual, great care was taken to ensure that the temperature was constant throughout the sample. The magnetic stirrer used to increase heat transfer was found to increase nucleation rates in supercooled water-in-oil emulsions, presumably by increasing the collision rate between droplets, so the cooling rate was set to $1°C\ h^{-1}$ to maximize the thermal

3.5 CRYSTAL NUCLEATION

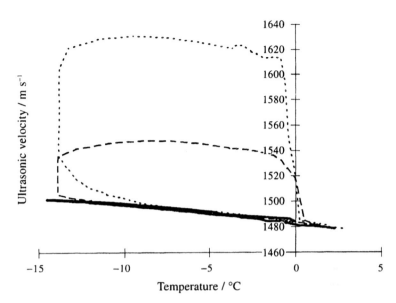

FIGURE 3.15 Variation of ultrasonic velocity with temperature for 20% v/v water-in-oil emulsions stabilized by lanolin in the presence and absence of ice nucleating agents. —, double distilled water; – –, water + AgI; · · · ·, water + SNOMAX (Archer *et al.*, 1996).

homogeneity of the sample. A dilatometer was fixed to the UVM cell to monitor density changes simultaneously, giving independent confirmation of phase changes in the water. Results are plotted in Figure 3.15. These results show very clearly that SNOMAX is a very effective ice nucleator, crystallizing 53% of the water, compared to 18% in the case of AgI and none in the absence of ice nucleators.

It is interesting and instructive to note that even gentle stirring with a magnetic flea had a very significant effect on the rate of nucleation of ice. In the absence of stirring, double distilled water did not crystallize during the course of the experiments. In the presence of gentle stirring, the majority of the water crystallized, presumably due to an increase in the droplet collision rate, thereby increasing the nucleation rate.

3.6 THE SOLUTION–EMULSION TRANSITION AND EMULSION INVERSION

There are practical reasons for being concerned about what happens to the velocity of sound when the particle size reduces in an emulsion. Colloid scientists work routinely with particles of a few hundred nanometers in size. Surfactant micelles are typically on the order of 10 nm in size, comparable to the largest proteins. The limit of particle size reduction is the elementary entity of which the particle is made up, a triglyceride molecule, for example. When analyzing solutions with ultrasound, we would like to know the relationship between ultrasound velocity and the properties of the molecules which make up solutions. There have been a number of empirical and semiempirical attempts to predict the ultrasonic properties of solutions from their composition. The method of Rao (Rao et al., 1966) is the best known, but there are many others (see §2.4.1.1). Scattering theory results from a later chapter will be used here to predict the ultrasound properties of microemulsions and inverted emulsions.

3.6.1 EMULSION INVERSION

The effects of emulsion inversion on ultrasound propagation has been considered by a number of authors (Ballaró et al., 1980; Tsouris and Tavlarides, 1994; Bonnet and Tavlarides, 1987) and there has been some debate about the appropriate model to use. The confusion has arisen because, from the point of view of the unmodified Urick equation, there will be no effect on ultrasound velocity when emulsion inversion occurs. However, once scattering is taken into account, effects may theoretically be expected due to two causes. First, it is unlikely that particle size will remain the same during emulsion inversion. If scattering is occurring, then a change may be expected in velocity due to changes in particle size and distribution. Second, it is shown in Chapter 4 and later in this chapter, that the effective compressibility contains a contribution from the thermal properties of the continuous and dispersed phases which is asymmetric in its form. In general terms, therefore, emulsion inversion may be expected to alter the velocity of sound in ways predicted by the modified Urick equation (Equation 2.34, §2.4.5) and the generalized scattering theory of Chapter 4.

An examination of Equation 2.34 shows that it is not symmetric with respect to interchange of the subscripts. This means that two systems which are identical in all respects, except for the fact that the continuous and dispersed phases exchange places, will have different ultrasonic velocities. The culprits are the term θ, which depends on the thermal expansivity, specific heat, temperature, and density of the two phases, and the scattering term $\frac{2}{3}(\Delta\rho)^2/\rho^2$.

In the context of a discussion of emulsion inversion, Equation 2.34 quantifies the effect of inversion on ultrasound velocity based on the firm ground of scattering theory.

3.7 DETERMINATION OF EMULSION STABILITY BY ULTRASOUND PROFILING

3.7.1 INTRODUCTION

Ultrasound profiling depends on differences in the density and compressibility between a dispersed phase and the continuous phase in a mixture. It is the changes in the concentration of the dispersed phase with height and time which can be detected in ultrasound velocity profiles.

As has already been seen (Figure 3.7), the relationships between the various components of a mixture can be complex. Also important is the difference in the temperature dependence of the properties of water and that of other components such as oil. In general it is essential to determine the temperature dependence of the velocity of sound, both in the mixture and in the continuous phase, using techniques explained in Chapter 2. An examination of Figure 3.7 shows that there is a temperature at which the velocity in the two phases will be equal. At this point ultrasound profiling will be very inaccurate, because of a lack of contrast between the properties of the components of the mixture. Thus, a correct choice of temperature for a profiling experiment is vital. The interpretation of ultrasound profiler data may be further complicated by the presence of ultrasound scattering. Techniques which reduce the impact of scattering on the calculated oil volume fraction against height profiles are presented later in this chapter (Section 3.7.4).

Ultrasound profiling involves the measurement of the velocity of sound in a column of liquid containing a dispersed material (See Figure 3.17). The differing density of the continuous and dispersed phase causes the emulsion to separate, by the dispersed phase either floating upwards (creaming) or settling downwards (sedimentation). The volume fraction of the dispersed phase at a particular height in the sample can be determined from the modified Urick equation (Equation 2.34), and the evolution of the volume distribution of the dispersed phase can then be plotted against height in the column and against time (Figure 3.16).

The creaming profile in Figure 3.16 illustrates a number of features which might be expected in a monodisperse (narrow size distribution) system of particles creaming in an unhindered way. In this example the oil-in-water dispersed phase is less dense than the continuous phase. First of all, the particles move up the column at a uniform rate, until they meet the top, where they stop. The formation of the cream could be seen clearly in the ultrasound data after 18 h, although it would not be visible to the eye until 43 h. As the particles move up they leave a clear layer at the bottom which is called serum. A sharp boundary appears between the serum and the emulsion, and this boundary moves up the tube to meet the cream at the top. The cream itself forms a boundary with the emulsion, and this moves down the tube as the cream fills up with oil drops from the emulsion. Creaming of this sort can be quite successfully modeled once hydrodynamic hindrance and thermal diffusion have been accounted for (Pinfield et al., 1994) and simpler models have been used to determine particle size distribution (Carter et al., 1986).

3.7.2 HISTORY

An early description of ultrasound profiling can be found in Attwood et al. (1981), who applied their technique to a study of diffusion of drugs in gels (Tolley and Rassing, 1983). The application of the technique to emulsions was developed at the Institute of Food Research, Norwich (Howe et al., 1985, 1986; Carter et al., 1986; Hibberd et al., 1986; Gunning et al., 1988, 1989; Howe and Robins, 1990;

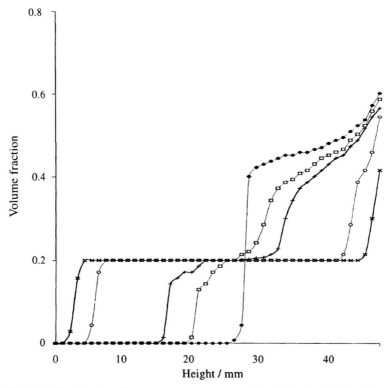

FIGURE 3.16 Creaming of *n*-tetradecane-in-water emulsion (15 wt% oil, 0.75 wt% protein, pH 7) containing 0.05 wt% xanthan. Volume fraction is plotted against height on storage at 20°C for ×, 18 h; ○, 43 h; +, 127 h; ☐, 154 h; ♦, 223 h. (From Cao *et al.*, 1988.)

Fillery-Travis *et al.*, 1992, 1993). Ultrasound profiling has been adopted elsewhere for the study of emulsion stability (Cao *et al.*, 1990, 1991; Dickinson *et al.*, 1993a; Wedlock *et al.*, 1990). A drawback of the method was the labor involved in measuring the time-of-flight of the ultrasound pulse, which was carried out on an oscilloscope by comparing the received pulse waveform with the transmitter pulse. Data reduction can also be time consuming because of the need to calibrate the system carefully. Other workers have automated the technique, an essential step if the technique is to gain wide acceptance (Wedlock *et al.*, 1993; Dickinson *et al.*, 1994; Pinfield *et al.*, 1994, 1995, 1996; Povey, 1995; Pinfield, 1996).

3.7.3 THE LEEDS PROFILER

The Leeds ultrasound profiler (Figures 3.17 and 3.18) was developed from the Institute of Food Research design, as modified by Dr. Wedlock at Shell Research, Sittingbourne, U.K. Many of the details of the system have been discussed in Chapter 2. A Hewlett-Packard Function Generator (Hewlett-Packard, U.K.)

3.7 DETERMINATION OF EMULSION STABILITY BY ULTRASOUND PROFILING

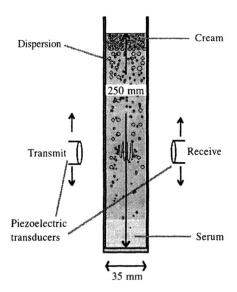

FIGURE 3.17 Schematic diagram of an ultrasound profiler.

generates a radio frequency tone burst with a 16-V output. The frequency of the tone burst is tuned to the resonant frequency of the transducer crystal. The apparatus comprises six glass tubes of cylindrical cross-section (250 × 35 × 35 mm). The tubes are held on a carousel which is driven by a stepper motor system under computer control. The transducers are 1.2-MHz medium damped probes manufactured by Sonatest Ltd. (Sonatest, U.K.), with an 18 mm diameter crystal. The transducers are driven up and down the outside of the glass tube by another stepper motor which is also under computer control. The trigger conditions affect the timing, and care must be taken to choose the appropriate trigger, depending on the signal conditions. For this reason, the trigger conditions are normally set by software and are preprogrammed. Good temperature control is crucial to success, so the glass tubes and associated sample changer are immersed in a thermostated water bath.

The sequence of positioning the transducers and the cells can be automatically controlled by the computer program. The program itself is written in Visual BASIC which makes it possible for anyone with a little basic training to use the apparatus. The program also takes care of acquiring data and storing it on Excel spreadsheets, where the data is automatically displayed in a graphical form. Calibration of the system is also automatic, using the distilled water calibration method explained in Chapter 2. In addition, the program warns if there are likely to be errors in the data. The program calculates the mean and standard deviation of 10 time-of-flight readings and takes the mean as the value at any given transducer position. Typically, standard deviations are 0.001 μs in 35 μs in water. A significant rise in the standard deviation is indicative of high attenuation, and high attenuation is a

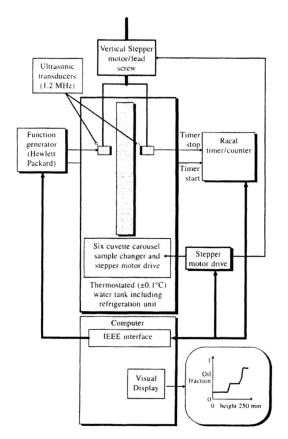

FIGURE 3.18 The Leeds ultrasound profiler.

sign that scattering and frequency-dependent effects may be a problem. In addition, high attenuation may give rise to triggering errors (see Chapter 2) which can give false readings. The software warns of this condition wherever possible. A jump in the timing of one or more periods of the tone burst is automatically detected by the software and corrected for (see Chapter 2). The correction is noted automatically on the spreadsheet. In addition, special methods have been developed for correcting for the particle-size and frequency-dependent effects associated with scattering (next section). More details of the computer control system are given in Chapter 5.

3.7.4 INTERPRETATION OF ULTRASOUND VELOCITY PROFILES

A method called renormalization has been developed to help account for the frequency-dependent and size-dependent effects associated with the scattering of ultrasound (Pinfield *et al.*, 1995, 1996). It is critical that these effects be correctly accounted for, because they can have a very significant effect on velocity (see Chapter 4). Two common approaches have been used in the past. The first is the Urick equation (Chapter 2), which uses volume-averaged values of compressibility and density in the Wood equation (Chapter 2) for the velocity of sound in a fluid. The second approach is multiple scattering theory, which will be considered in detail in the next chapter. Multiple scattering theory often requires numerical solution and very many physical properties of the constituent phases (see Chapter 4 and the Appendix). This is impractical for everyday use in a laboratory apparatus. It has been shown (Pinfield et al., 1995) that a modified Urick equation (Equation 2.34) can be written which is a special case of multiple scattering theory, in the limit of low frequency. It is usually not known whether the Urick equation or scattering theory results are valid, because of a lack of the necessary data. However, Equation 2.34, §2.4.5, may be thought of as a description of the behavior of ultrasound velocity where the coefficients α and δ are unknowns, to be determined experimentally. In this case the scattering coefficients are, in effect, being determined experimentally (§4.3.10.2).

3.7.4.1 Renormalization

The renormalization method (Pinfield et al., 1996) enables concentration profiles to be determined from velocity measurements with a knowledge only of the continuous phase velocity and the initial volume fraction. This accounts for the effects of scattering not included in the modified Urick equation (Equation 2.34). In the absence of information about the initial volume fraction, relative changes in volume fraction can be determined.

The basis of the method is the conservation of volume of the dispersed phase. This assumes that there is no compression of the dispersed phase droplets during creaming. It also assumes that, in the case when the velocity of sound depends on particle size, the particle size distribution does not change significantly during the course of an experiment. If the initial volume fraction ϕ_0 is known, then the volume fraction, averaged over the full height of the sample, must be constrained by

$$\langle \phi \rangle_{h,t} = \phi_0, \tag{3.11}$$

where $\langle \ \rangle$ represents the average over the height h of the sample, of the scalar quantity contained within them. t is the elapsed time in the experiment, at which the scan over the height h was taken. Equation 3.11 can be rewritten in terms of velocity as

$$\phi = a\left(\frac{1}{v^2} - \frac{1}{v_1^2}\right) + b\left(\frac{1}{v^2} - \frac{1}{v_1^2}\right)^2, \tag{3.12}$$

where

$$a = \frac{1}{a} \quad \text{and} \quad b = \frac{-\beta}{\alpha 3}. \tag{3.13}$$

α and δ are defined in Equation 2.34, §2.4.5. For convenience, we write

$$\Delta\left(\frac{1}{v^2}\right) = \left(\frac{1}{v^2} - \frac{1}{v_1^2}\right). \tag{3.14}$$

The coefficients a and b can be calculated from Equations 3.11 and 3.12, on each scan of the sample height. On the first scan, the second-order terms may be neglected, providing no significant creaming has taken place, and the linear term may be estimated from

$$a = \frac{\phi_0}{\Delta\left(\frac{1}{v^2}\right)_{h,0}}. \tag{3.15}$$

A linearly renormalized profile can then be determined from

$$\phi = a.\Delta\left(\frac{1}{v^2}\right). \tag{3.16}$$

It is easy to determine if this renormalization is sufficient on later scans because Equation 3.11 will hold, that is, the linear renormalization will predict the total volume fraction correctly. If it does not, then the quadratic coefficient must be calculated:

$$b = \frac{\phi_0 - a\langle\Delta(1/v^2)\rangle_{h,t}}{\langle(\Delta(1/v^2))^2\rangle_{h,t}} = \frac{\phi_0}{\langle(\Delta(1/v^2))^2\rangle_{h,t}} \times \left(1 - \frac{\langle\Delta(1/v^2)\rangle_{h,t}}{\langle\Delta(1/v^2)\rangle_{h,0}}\right). \tag{3.17}$$

Once b has been calculated, the volume fraction can be obtained from Equation 3.12 for all scans. Estimates of a and b may be improved by using data from later scans, because the accuracy of the quadratic term will improve as the sample becomes more nonuniform.

Even if the quadratic term is significant on the first scan, values for a and b can be obtained from two different scans:

$$a = \frac{\left[\dfrac{\phi_0}{\langle\Delta(1/v)^2\rangle_{h,0}}\right] \times \left(1 - \dfrac{\langle(\Delta(1/v)^2)^2\rangle_{h,0}}{\langle(\Delta(1/v)^2)^2\rangle_{h,t}}\right)}{\left(1 - \dfrac{\langle(\Delta(1/v)^2)^2\rangle_{h,0}}{\langle(\Delta(1/v)^2)^2\rangle_{h,t}} \times \dfrac{\langle\Delta(1/v)^2\rangle_{h,t}}{\langle\Delta(1/v)^2\rangle_{h,0}}\right)} \tag{3.18}$$

$$b = \frac{\left[\dfrac{\phi_0}{\left\langle\left(\Delta(1/v^2)\right)^2\right\rangle_{h,0}}\right] \times \left(1 - \dfrac{\left\langle\left(\Delta(1/v^2)\right)\right\rangle_{h,0}}{\left\langle\left(\Delta(1/v^2)\right)\right\rangle_{h,t}}\right)}{\left(1 - \dfrac{\left\langle\left(\Delta(1/v^2)\right)^2\right\rangle_{h,t}}{\left\langle\left(\Delta(1/v^2)\right)^2\right\rangle_{h,0}} \times \dfrac{\left\langle\Delta(1/v^2)\right\rangle_{h,0}}{\left\langle\Delta(1/v^2)\right\rangle_{h,t}}\right)}. \quad (3.19)$$

Care must be taken with Equations 3.18 and 3.19 to ensure that the values obtained for the terms in these equations are significant relative to experimental errors in velocity measurements. In most cases, Equations 3.15 and 3.17 will give as good or better results.

3.7.4.2 Limits of Applicability of Renormalization Method

Equation 3.12 assumes that the attenuation is insignificant relative to the real part of the wavenumber (see §2.3). This is normally the case because the equipment is incapable of measuring samples when the attenuation is so large that the imaginary part of the wavenumber (see §2.3) $k'' \geq k'$, the real part of the wavenumber. The dominant restriction is the requirement that any change in the velocity caused by changes in particle size distribution in the sample be negligible. Creaming in polydisperse samples is characterized by fractionation due to differential creaming rates. This can lead to significant errors in volume fraction at particular points in the cream. However, provided the particle size distribution in the sample is small enough in comparison to the error in measuring velocity, the renormalization method will still be applicable. This condition may be summarized by the following equation:

$$\left| v(\phi(r_{max})) - v(\phi(r_{min})) \right| \leq \Delta v_{exp}. \quad (3.20)$$

where $v(\phi(r_{max}))$ and $v(\phi(r_{min}))$ are the ultrasound velocity in the regions of largest and smallest average particle size, respectively. The difference must be smaller than the experimental error Δv_{exp}. Obviously, this condition is satisfied for a monodisperse sample, but a high degree of polydispersity may also be tolerable if the velocity is independent of particle size over the size range in the sample. In the long-wavelength limit, broad variations of particle size may cause little change in ultrasound velocity; a more detailed discussion of this appears in Chapter 4.

In summary, the volume fraction in an ultrasound profiler can be determined from the velocity under a wide range of conditions, provided the renormalization method is used as just described. Only the continuous phase velocity need be known in advance, but if the initial volume fraction is known, then the method will determine the volume fraction throughout the experiment.

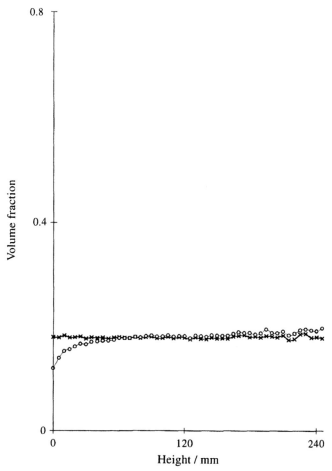

FIGURE 3.19 Plot of volume fraction against height for a creaming oil-in-water emulsion comprising 18 vol.% mineral oil and 2 wt% Tween 20 at 30°C. $d_{32} = 0.65$ µm. Scans after ×, 3 h; ○, 188 h.

3.7.5 EXAMPLES OF PROFILING

Oil-in-water emulsions prepared from sunflower and mineral oils, together with the small-molecule amphiphilic surfactant Tween 20 or the protein emulsifier sodium caseinate, have been extensively studied with the ultrasound profiler. A remarkable range of behavior has been observed from apparently similar systems (Povey, 1996).

In Figure 3.19 a stable oil-in-water emulsion is shown. To the eye, no change is visible, but a reduction of 8% in the oil concentration at the bottom of the tube is apparent from the ultrasound data. When a very small amount of the microbial polysaccharide xanthan is added, the situation changes dramatically (Figure 3.20). Flocculation occurs, and the flocs cream more rapidly because of a net reduction in the drag resulting from the motion of liquid around them. A distinct serum layer is visible by eye, but the boundary between the serum and the emulsion does not then move. Computer modeling experiments (Pinfield *et al.*, 1994) indicate that this boundary either will move up the tube, or will not be apparent in a nonflocculated emulsion. The absence of any cream layer is suggestive that flocculation has led to the formation of a weak gel in the emulsion.

Figure 3.21 shows observed effects which are in some ways similar to those of Figure 3.16, but in this case it is likely that a large particle diffusion component blurs out the serum/emulsion boundary. A further feature of these profiles is that the cream layer is undergoing compression, leading to an increase in concentration. By "compression" it is meant that the particles rearrange to get a higher concentration. "Compression" in this context does not mean that the particles are squashed to give a density increase.

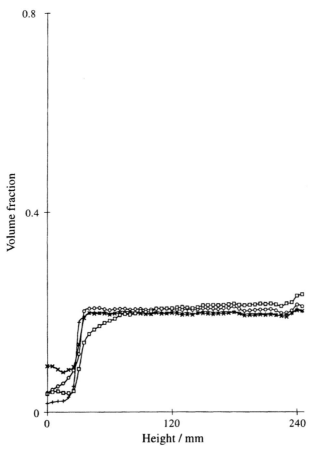

FIGURE 3.20 As in Figure 3.19 except with 0.017 wt% xanthan added to the continuous phase: ×, 1 h; ○, 9 h; +, 19 h; □, 138 h (Dickinson *et al.*, 1994).

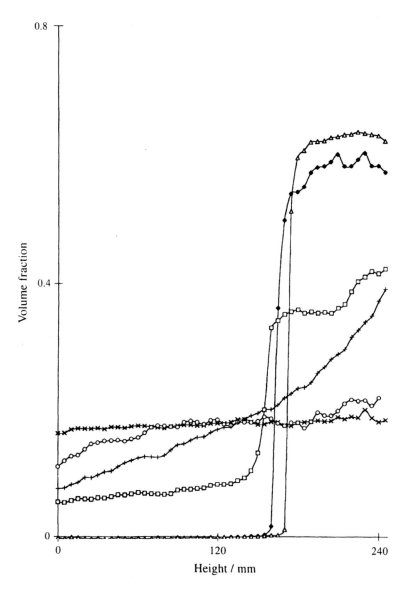

FIGURE 3.21 Figure 3.19 except with 0.173 wt% xanthan added to the continuous phase: ×, after 11 h; ○, after 20 h; +, after 30 h; ☐, after 45 h; ♦, after 99 h; Δ, after 188 h (Dickinson *et al.*, 1994).

The profile in Figure 3.22 has a static boundary between the serum and the emulsion, a characteristic it shares with Figure 3.20. It might therefore be assumed that the emulsion has gelled. But the development of a cream layer indicates that gel development was delayed, at least until part of the oil had formed a cream.

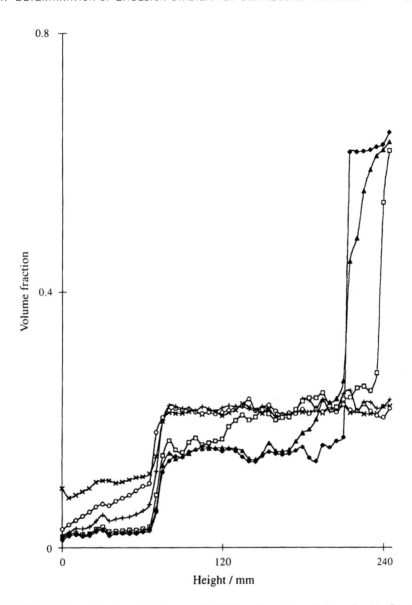

FIGURE 3.22 As in Figure 3.19 except with 0.035 wt% xanthan added: ×, after 1 h; ○, after 6 h; +, after 10 h; □, after 19 h; ◆, after 29 h; Δ, after 189 h (Dickinson *et al.*, 1994).

This is consistent with the reduced volume fraction of oil in the middle of the emulsion at the end of the experiment. This profile is interesting because it appears that the processes of flocculation, creaming, and gel formation are taking place together within the same system. In contrast, in Figure 3.23 a lower xanthan concentration has been sufficient to destabilise the emulsion, but is apparently insufficient to gel the emulsion.

The profiles in Figure 3.24 are different from the rest of the profiles in this section in that they comprise a much higher oil volume fraction and a different stabilizer. Here, a serum phase appears which grows in volume, while the development of a cream has been arrested. Rheological studies indicate that the emulsion phase has gelled in these experiments, and it can be inferred that polymer

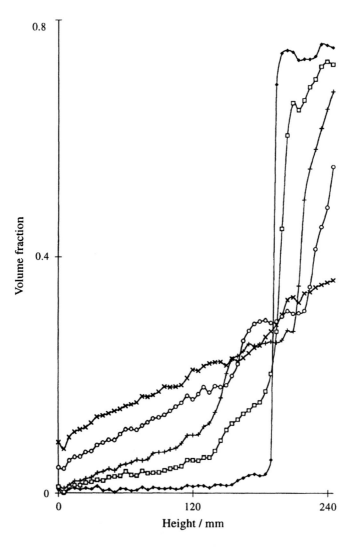

FIGURE 3.23 As in Figure 3.19 except with 0.069 wt% xanthan added: ×, after 1.5 h; ○, after 4 h; +, after 6 h; □, after 10 h; ♦, after 19 h (Dickinson et al., 1994).

is being displaced from the gel into the serum, which grows as a consequence. Perhaps the gel phase is concentrating under its own weight in this system, as has been suggested may occur by other workers (Parker *et al.*, 1995). This process is called "syneresis." Ultrasound profiling may therefore be used to study syneresis in a quantitative manner.

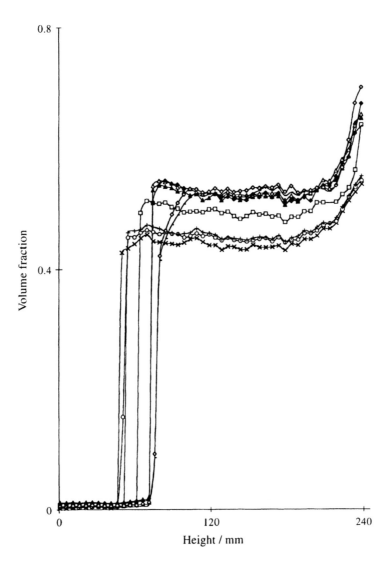

FIGURE 3.24 Volume fraction plotted against height for a 35 wt% *n*-tetradecane oil-in-water emulsion containing 4 wt% sodium caseinate, 30°C, pH 6.8, d_{32} = 0.5 µm: ×, after 52 h; ○, after 64 h; +, after 3 d; □, after 6 d; ▲, after 12 d; ∆, after 15 d; ♦ after 19 d; ◊, after 33 d (Dickinson, Golding, and Povey, 1996).

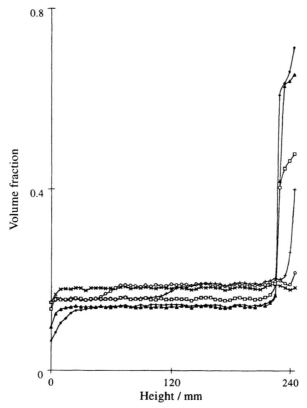

FIGURE 3.25 As in Figure 3.19 except with 0.05 wt% xanthan and 0.05 M NaCl added: ×, after 1 h; ○, after 3 h; +, after 5 h; □, after 8 h; ▲, after 43 h; ♦, after 138 h (Dickinson *et al.*, 1994).

Finally, the profile in Figure 3.25 exhibits no significant serum, but does show the beginnings of the development of a cream.

3.8 SUMMARY

Ultrasound velocity measurement is a powerful tool for characterizing multiphase fluid media. It can be used for composition monitoring and the study of changes in phase, for example, in water and oils. It is an effective method for determining solid fat content. Creaming profiles such as the ones shown in this chapter can give detailed indirect information about the state of flocculation of an emulsion. The ultrasound profiling technique is a valuable tool for characterizing emulsions and for studying shelf life and stability. The examples in this chapter strongly suggest a far wider scope for the application of ultrasound than is presently the case. In the next chapter, techniques which give more direct information about the state of aggregation of an emulsion are considered.

4

SCATTERING OF SOUND

4.1 THEORIES OF SOUND

The Wood equation (Equation 2.8), which has the virtue of simplicity, applies to pure liquids. But very few liquids of interest are pure, and in general the velocity of sound is a complicated function of a large number of physical, chemical, and biochemical properties. There are a number of reasons that acoustic propagation is complicated. First of all, the displacements of the elementary entities (particles) which constitute a material are related to the restoring forces on the particle, which depends on the surrounding particles. This introduces questions of symmetry, particle–particle separations, and interparticle forces and their dependence on separation. Second, sound waves have temperature waves associated with them. It is only above about 10^9 Hz in air and 10^{12} Hz in water that propagation becomes isothermal (Chapter 1). At all frequencies considered in this book, acoustic propagation in pure liquids is a very good approximation to adiabatic. Although the amplitude of the temperature wave is only a few millikelvins (see Table 2.1), this wave can have a profound and subtle influence on acoustical propagation, as will be seen. Third, changes in elastic and density properties due to inhomogeneities in the medium cause reflection, refraction, scattering, diffraction, and interference in the propagating wavefront. Finally, density differences give rise to scattering in fluids due to relative motion of inhomogeneities with respect to the continuum. This scattering is related to the viscosity of the continuum.

The Urick equation (Chapter 2) is an explicit statement of a principle implied by the Wood equation, which is that the suspension is considered to behave like a homogeneous fluid, with a modified compressibility and density corresponding to the volume-averaged values for the suspension. It does not consider the interaction of the sound wave with the individual particles. This is called a "homogeneous description." We will show here that the volume averaging of compressibility and density, combined into the Urick equation (Equation 2.9) is best viewed as a special case of scattering theory.

It would take a far longer work than this one to deal with the entire subject of acoustic propagation. The practical objective here is to provide the means to analyze velocity of sound data in fluids under conditions accessible to modern commercially available equipment. To achieve this it is necessary to bring theoretical implementations within the capabilities of personal computers.

There are a number of solutions to acoustic propagation in dispersed systems and the reader is referred to Harker and Temple (1988, 1992) for a comprehensive review of the subject and the relative merits of the various approaches. Meeten and Sherman (1993) attempted to combine the viscous and elastic effects in suspension; they fitted explicit formulas to experimental velocity and attenuation data using estimates of the shear-dependent viscosity. This area awaits further development; simple explicit formulas for propagation in viscoelastic media are given in Chapter 5. Empirical theories are considered briefly in §2.4.1.1.

Velocity measurement has been an underused probe of scattering and has several advantages over attenuation measurement. First, the effects of scattering can be much more accurately and sensitively probed with sound velocity. This is because it is a simple matter to determine the group velocity of a pulse to within parts in 10,000, whereas accuracy of a few percent is hard to obtain when measuring the acoustic attenuation of materials. In addition, sound velocity is much less sensitive to the approximations fundamental to the weak scattering treatment. This is because the phase shifts in the acoustic wave, which are detected in the form of sound velocity changes, are insensitive to changes in amplitude in the scattered wave. Sound velocity is also, for the same reason, more accurately described by theories of multiple scattering. Nevertheless, there are valid reasons for wanting to measure attenuation, and some of these will be discussed in the section on particle sizing.

4.2 A COMPARISON OF ELECTROMAGNETIC AND ACOUSTIC PROPAGATION

A comparison of optical and acoustic descriptions of wave propagation in materials appears in Table 4.1. There are a great many similarities between electromagnetic and acoustic scattering, as has already been pointed out (Chapter 1, §1.5), and this is particularly so at the level of the formal mathematical apparatus involved—so much so that entire books have been devoted to the explication of scattering theory for the two cases (Jones, 1986; DeHoop, 1995). Neither of these large books actually answers the practical questions posed here, and one of them neglects the effects of thermal waves.

4.3 SCATTERING THEORY

TABLE 4.1 Analogy between Electromagnetic and Acoustic Wave Propagation[a]

Electromagnetic waves	Acoustic waves
The velocity of propagation, c, is $$c^2 = \frac{1}{\varepsilon\mu},$$ where ε is permittivity and μ is permeability.	The velocity of propagation, v, is $$v^2 = \frac{1}{k\rho},$$ where k is complex compressibility and ρ is density.
Assume μ = *constant*; then $\varepsilon = \varepsilon' + i\varepsilon''$.	Assume ρ = *constant*; then $k = \kappa' + i\kappa''$.
Define the wave vector, $\mathbf{k} = k' + i\alpha$, $$c^2 = \frac{\omega^2}{\mathbf{k}^2}.$$	Define the wave vector, $\mathbf{k} = k' + i\alpha$, $$v^2 = \frac{\omega^2}{\mathbf{k}^2}.$$
Comparing real and imaginary parts gives $$\varepsilon' = \frac{1}{\mu\omega^2}\left[k'^2 - \alpha^2\right]$$ and $$\varepsilon'' = \frac{2k'\alpha}{\mu\omega^2}.$$	Comparing real and imaginary parts gives $$\kappa' = \frac{1}{\rho\omega^2}\left[k'^2 - \alpha^2\right]$$ and $$k'' = \frac{2k'\alpha}{\rho\omega^2}.$$

[a]The symbols used for electromagnetic propagation are used for this purpose only in this table and have a different meaning elsewhere in this work.

A helpful analogy has been drawn between electromagnetic and acoustic wave propagation by Evans (1986). Although this analogy should not be pursued too far, it indicates a correspondence between the permittivity of the medium for electromagnetic waves and the compressibility for acoustic waves. These analogies are the result of the common wave nature of the two disturbances. In particular, the wave equation (Equation 4.1) is the same for both forms of radiation. It suggests that a complex wave velocity can be defined for both cases, which can be particularly helpful in transferring results from the electromagnetic case to the acoustic case, and vice versa. For example, Rayleigh scattering, in which the scattering intensity depends on the fourth power of the wavelength, occurs both in optics (producing the blue sky) and in acoustics (Strutt, 1872). The example of the quarter wave-transformer was noted in Chapter 3.

4.3 SCATTERING THEORY

4.3.1 WHY SCATTERING THEORY?

In this work, scattering theory is the chosen solution to the acoustic propagation problem, primarily because it is very well suited to the analysis of acoustic propagation in dispersions in the long wavelength limit (acoustic wavelength $\lambda \gg$ particle diameter d). The theory itself provides an understanding of the physical processes which occur at the particles suspended in liquid media. It also successfully combines the effects of particle displacement and heat flow into a single description of propagation. Scattering theory provides a firm quantitative base for analyzing ultrasound propagation. With it, it is possible to size particles, determine volume fraction, and follow the evolution of aggregation of particles. It is essential to grasp the nettle of scattering theory if practical progress is to be made in these important areas.

The apparatus of scattering theory has been of great practical benefit. Computer models have aided the interpretation of ultrasound data from complex creaming experiments (Pinfield et al., 1996). The ability to calculate ultrasound parameters for solid and liquid particles has permitted the determination of crystallization kinetics (Dickinson et al., 1993b). The size of oil particles dispersed in mayonnaise was determined without dilution (McClements et al., 1990c). Scattering theory forms the basis for particle sizing apparatus now in commercial development (Alba, 1992; Roberts, 1996).

Sound propagation through any mixture involves scattering of sound. It is commonly thought that scattering solely involves changes in acoustic amplitude and is often considered synonymous with backscattering. It is often forgotten that scattering involves amplitude changes in all directions, including the forward direction, and includes both increases and decreases in amplitude. Even more startling is the fact that scattering can occur without any amplitude changes whatsoever, in any direction. This is because the phase of the signal can be changed by scattering, and this is commonly observed as a change in the velocity of sound. Looked at another way, sound velocity changes in materials may be viewed as the result of scattering.

In this book, media which support the shear mode of sound propagation over macroscopic distances will not be considered. Thus, most solid materials are excluded from the discussion, except for their role as scattering particles. Soft solids and viscoelastic materials, however, are included because they do not support shear propagation as a bulk mode; they will be discussed later in this chapter.

The discussion on scattering which follows borrows heavily from the discussion on ultrasound propagation in Pinfield (1996).

4.3.2 WHAT IS SCATTERING? ASSUMPTIONS OF SCATTERING THEORY

Propagation in inhomogeneous media always involves scattering. Any process which transforms energy in one form or mode of motion and transforms it into another is called *scattering*. This definition does not require any energy to be dissipated from the wave motion, nor does it require the amplitude of the wave motion to change, since changes in the phase of the wave are an integral part of the scattering process (Figure 4.1).

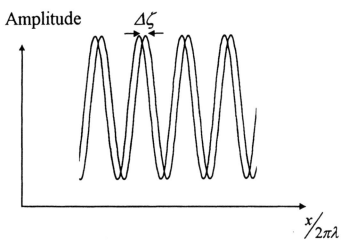

FIGURE 4.1   Phase difference between two sine waves.

The phase (ζ) of a wave defines its position in space, relative to some arbitrary point. Phase is therefore only meaningful when measured relative to a defined point.

$$\Delta p = \Delta p_0 \exp\{i(\omega t - \zeta + i\alpha x)\} = \Delta p_0 \exp(i\omega t)\exp(-\zeta)\exp(i\alpha x), \quad (4.1)$$

where

$$\zeta = k'x, \quad k' = \frac{\omega}{v_p}, \quad \text{and} \quad \mathbf{k} = k' + ik'' = k' + i\alpha.$$

Here Δp and Δp_0 are the instantaneous pressure deviation and the maximum pressure deviation, respectively; $i = \sqrt{-1}$; ω is the radial frequency; k' and k'' are the real and imaginary parts of the complex wave vector $\mathbf{k}$; v_p is the phase velocity of the wave (§2.1.2.7); and α is the attenuation coefficient.

Of course, dissipation may, and normally does, accompany scattering. The point is that scattering may occur without any dissipation of the total energy of the wave. This is especially relevant if *forward scattering* is to be properly understood. Here the wave is scattered into the forward direction, which can give rise to significant changes in the velocity of sound, without a correspondingly large change in signal amplitude or dissipation. In the dispersed systems which are the subject of much of this work, however, energy is scattered in all directions and removed from the forward direction. The detected wave is the sum of that part of the wave which is not scattered together with the forward-scattered waves. The summation of the direct and forward-scattered waves produces a phase shift which is interpreted as a change in the phase velocity of the wave (see §4.3.3.5).

It is possible for two wave motions to be coupled together to form a coupled wave. This is the case in sound, where particle displacement and temperature displacement go together (Figure 4.2). An increase in pressure causes the temperature to rise (except in the case of water below 4°C, where it will fall). On the expansion phase of the wave, the opposite occurs. This temperature fluctuation becomes significant at a discontinuity in the medium, where sudden changes in the thermal properties give rise to a discontinuity in the temperature. This, in turn, generates a scattered pressure wave, giving rise to "thermal scattering" (see §4.3.3.3).

The basis for scattering theory was laid in the 19th century by Lord Rayleigh (Strutt, 1872, 1896). The underlying idea is that of *partial wave analysis* (PWA). PWA involves writing the equation of any acoustic wave as a linear combination of different waves, each of a single frequency. Then only one frequency need be considered, since the results can be superposed. The time dependence of each wave can easily be dealt with through the term $\exp(i\omega t)$ (see Equation 4.1). Derivatives with respect to time are then simply obtained by multiplying by $i\omega$. A second PWA involves the spatial variation of the wave and its decomposition as the sum of different modes (i.e., monopole, dipole, etc.).

The work of Rayleigh has been built on over the years by a number of workers. Important historical landmarks in the application of partial wave analysis to acoustical scattering include the calculation of the scattering coefficients of a fluid droplet suspended in a fluid continuum (Epstein and Carhart, 1953), the extension to the determination of velocity and attenuation using multiple scattering theory (Lloyd and Berry, 1967), and the experimental vindication of scattering theory in a model emulsion (hexadecane in water) and a model suspension (polystyrene spheres in water) by Allegra and Hawley (1972). Other workers have considered the effects of a shell-like coating (Anson and Chivers, 1993; Sayers, 1984).

In the scattering theory method, a single spherical object is imagined to be suspended in an infinite liquid medium through which a plane wave of single

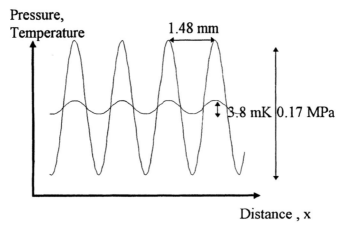

FIGURE 4.2 Pressure varying with position with the associated temperature variation, figures quoted for an acoustic intensity of 10 kW m^{-2} and 1 MHz in water. Similar curves may be drawn for the variation in time.

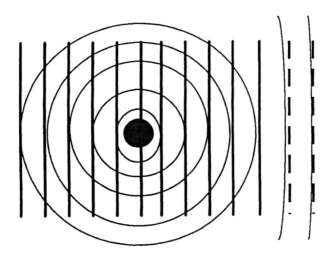

FIGURE 4.3 Schematic of spherically symmetric scattered waves, tending toward plane at an infinite distance.

frequency is propagating. The waves scattered at the surface of the object, regardless of their initial shape, will tend towards plane at an infinite distance away from the object (Figure 4.3). The receiver of the wave, within which both the directly transmitted and the scattered waves are combined, is considered to be infinite in extent, situated an infinite distance away. The total effect of a number of scatterers is then obtained simply by adding their combined effects together. At infinity the phases of the forward-scattered wave all combine, while the phases of waves scattered in other directions will cancel out, provided the scatterers are distributed randomly. Thus, we have another assumption—that of a random distribution of scatterers.

Because of the spherical nature of the scatterer and hence sphericity of the boundary conditions, the scattered wave potential (§4.3.3) may be expanded in spherical harmonics, and appropriate radial and angular functions are required to do this. These are called spherical Bessel functions (radial) and associated Legendre polynomials (angular). These functions form infinite series which can normally be truncated after two terms (dipole scattering) to obtain an accurate approximation in the cases considered here.

The scattered wave can be rescattered in a process called *multiple scattering* (see Figure 4.4). Since the intensity of the scattered wave is greatly reduced each time it is scattered, multiple scattering theory need only take one or two scattering events into account under conditions of weak scattering (§4.3.5.2). Multiple scattering may be thought of as a sequential series of scattering events. A scattered wave leaves one particle and hits another (2 in Figure 4.4), thus producing the second wave. It is a little difficult to reconcile this scheme with the fact that each "partial wave" lasts for an infinite time, and it is more accurate to consider the acoustic field

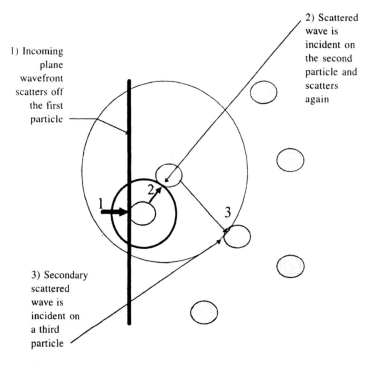

FIGURE 4.4   Multiple scattering of a plane wave by several particles.

incident at the particle as some sort of average of the contributions from the surrounding particles and from the unscattered part of the wave.

The two major elements of the scattering theory technique comprise (a) determination of the effect of a planar wave on a single isolated particle and (b) determination of the effect of an ensemble of particles in terms of the scattering properties of the individual particles.

The solution of any complex problem requires some degree of simplification. Scattering theory is no exception. A detailed list of assumptions and approximations follows (some of these are taken from Epstein and Carhart, 1953, and others are taken from Pinfield, 1996):

- Long wavelength limit (sometimes called the Rayleigh limit, after Lord Rayleigh; Strutt, 1872, 1896)
- Small attenuation in the continuous phase
- Plane incident wave
- Weak scattering
- Randomly distributed scatterers
- Adiabatic approximation
- Navier–Stokes form of the momentum equation
- Thermal stresses neglected

4.3 SCATTERING THEORY

- No changes in phase (vaporization or crystallization)
- Equations linearized with respect to the oscillatory variations of the field quantities (velocity, pressure, and so on)
- Temperature variation of the viscosity and heat conduction is neglected
- Gradual changes in temperature and pressure caused by the wave are neglected
- System is static
- Particles are spherical
- Infinite time irradiation
- Pointlike particles
- No overlap of thermal and shear waves
- No interactions between particles
- Lack of self-consistency

Despite this impressively long list of assumptions and approximations, scattering theory works extremely well for propagation in liquids. The following sections, which draw heavily on the discussion of scattering in Pinfield (1996), discuss the most important limitations imposed by the listed assumption.

4.3.2.1 Long Wavelength Limit

One significant simplification is to assume that the wavelength of sound in the fluid is much greater than the size of the inhomogeneities within the fluid. This is not a very restrictive condition, although it does exclude the consideration of suspended objects whose size is of the order of, or greater than, the wavelength of sound (1.48 mm in water at 20°C and 1 MHz). For comparison, colloidal systems have particle sizes below about 5 μm and water-in-oil emulsions may contain water droplets of up to 0.1 mm in size. The long wavelength limit is more likely to break down in colloidal systems as a result of increasing frequency because for wavelength $\lambda \leq 0.3$ μm the frequency $f \geq 50$ MHz.

The long wavelength limit is defined in terms of the wave vector k and the particle radius r as

$$kr \ll 1, \tag{4.2}$$

or in terms of the wavelength,

$$\frac{2\pi r}{\lambda} < \frac{1}{10}, \tag{4.3}$$

or in terms of the parameter $f^{0.5}r$,

$$f^{0.5}r < \sqrt{\frac{v}{20\pi}} \times \sqrt{r}, \tag{4.4}$$

The significance of the parameter $f^{0.5}r$ is that it is proportional to the thermal and shear wavelengths in liquids. In the long wavelength limit, the ultrasound velocity and the attenuation per wavelength ($\alpha\lambda$) vary with $f^{0.5}r$.

If wavelength is similar to or less than the diameter of the scatterer then we have geometric scattering. The mathematical treatment of this region is much more complex and it is much more difficult to make general statements about the scattering in these circumstances. So the question is avoided in this book on the basis that sound velocity data reduction for data from geometric scattering is much less amenable to analysis.

4.3.2.2 Low Attenuation

This is the same as the condition discussed in §2.3, that is, that $k' \gg k''$. In other words, the real part of the wavenumber should be much greater than its imaginary part, which is the attenuation α ($k' \gg \alpha$).

4.3.2.3 Plane Wave

The effects of a lack of planarity on acoustic measurements are discussed in Chapter 2. In an accurate acoustic system, it is important to ensure a high degree of planarity in the wave, so this condition is generally met in well-designed apparatus.

4.3.2.4 Scattering Is Weak

The condition that the amplitude of the scattered wave is far less than the amplitude of the incoming wave is called *weak scattering*. This is not so restrictive a condition and really only excludes the consideration of resonant scattering from bubbles in water, which will be considered separately.

4.3.2.5 Random Distribution of Particles

There are good theoretical reasons for believing that an ordered distribution of particles can make a significant difference to the scattering. For example, an assumption implicit in the PWA treatment is that the scatterers are distributed randomly, so a nonrandom distribution should alter the scattering. Scattering theories normally account for the systematic distribution of scatterers through the introduction of a structure factor. The structure factor is studied by diffraction methods where the amplitude of the scattered signal is measured as a function of the scattering angle, as for example in X-ray or neutron diffraction. In this work we consider primarily the forward-scattered wave and the structure factor will be considered no further. This is not to say that it is unimportant, but we have little data on acoustical structure factors.

4.3.2.6 Adiabatic Approximation

In Chapter 1 it was shown that acoustic propagation in pure liquids is adiabatic under all conditions considered in this work. However, the situation in a mixture of liquids or of solid in liquid is perhaps not so clear. This is an important question which has been referred to indirectly in Chapter 2. It will be shown later in this chapter that the modified Urick equation (Equation 2.34) is a special case of scattering theory, and it is recommended that this equation be used to determine the adiabatic compressibility from sound velocity and density measurements. Other workers have used the Wood equation to determine adiabatic compressibility. Zemansky derives the condition for adiabatic propagation in pure liquids as

$$\frac{2\tau f}{v^2 \rho C_V} \ll 1, \quad (4.5)$$

where τ is thermal conductivity; v, velocity of sound; ρ, density; C_V, specific heat at constant volume; and f, frequency of sound. The left-hand side of this equation has the value 1.3×10^{-7} at 20°C for 1 MHz ultrasound in water.

The condition for adiabaticity, as defined by Pierce (1981b) and Zemansky (1957c), depends on the absence of heat flow from the compression half cycle to the expansion half cycle of the acoustic wave (see Pinfield and Povey, 1997). Heat exchange between solute and solution extends over distances of the order of the thermal wave decay length, which, for a particle in water, is given by $\sqrt{2\tau/\rho C_p \omega}$ (= 0.2 mm at 1 MHz), where ω is the radial frequency of the acoustic wave (s^{-1}); r, the density (kg m^{-3}); τ, the thermal conductivity (J m^{-1} K^{-1}) and C_p, the specific heat capacity at constant pressure (J kg^{-1} K^{-1}).

For liquids such as water, the right-hand side of this condition is less than 10^{-6} m s^{-1}. This condition (Equation 4.5) is the same as that imposed by Epstein and Carhart (1953), who pointed out that the possibility of separating the attenuation (and by implication the propagation equations themselves) into a viscous and a thermal part is entirely due to the smallness of the ratio

$$\frac{|k|}{|k_T|} = \frac{2\tau f}{\rho C_p v^2}, \quad (4.6)$$

where k_T is the thermal wavenumber. This is, in fact, the same ratio as Equation 4.5 and shows that Zemansky's condition amounts to demanding that the acoustic wavelength should be much greater than the decay length of the thermal wave. This quantity reappears often in the scattering equations.

It may be concluded that propagation is very close to adiabatic under the conditions pertaining in the experiments considered in this work, provided scattering remains weak. This will be discussed later. Finally, it is not the case that conditions are *exactly* adiabatic. There is dissipation associated with the scattered thermal waves; the point is that it is so small that its effect on velocity through the complex part of the wave vector can be ignored. This is a subtle point, because the thermal properties of the system can considerably affect the velocity and these effects have often been ignored. This will be discussed later.

4.3.2.7 Navier–Stoke's Form for the Momentum Equation

The use of the Navier–Stokes form for the momentum equation is a feature of the way the scattering equations have been solved, rather than an absolute limitation of scattering theory. It is perfectly feasible to include other hydrodynamic terms, which can account for various types of non-Newtonian flow, including a small rigidity term, providing the shear mode in the continuous phase remains small and the equations can be linearized. This condition may be restated as requiring that the shear mode remains insignificant.

4.3.2.8 Thermal Stresses Neglected

Thermal stresses arise because of dimensional changes associated with temperature fluctuations caused by the passage of the sound wave. Their neglect is not likely to have a serious effect, but their exact form has not been calculated (Epstein and Carhart, 1953).

4.3.2.9 No Changes in Phase

The effects of changes in the phase of a dispersion, through melting, solidification, or vaporization, have already been discussed in Chapter 3. It remains only to note here that these effects can be very large and can completely invalidate scattering theory when they occur.

4.3.2.10 Linearization of Equations

The question of linearity has been addressed in §2.1.2.6, where it was shown that under most conditions pertaining in ultrasound velocity measurements, propagation was well within the linear approximation.

4.3.2.11 Temperature Variations

At the power levels holding in ultrasound velocity measurements (§2.1.2.6), the rise in temperature due to thermal dissipation (heat) associated with the passage of the sound wave may be ignored. However, this is not necessarily the case when phase changes are occurring (see §4.3.2.9 earlier).

4.3.2.12 System Is Static

Each particle is assumed not to move during the passage of the sound wave, other than because of the acoustic field. The characteristic times for diffusion and creaming (the dominant kinetic processes in a colloidal dispersion) are normally much longer than the duration of the ultrasound pulse (a few microseconds). Therefore, the particles may be considered stationary. This may not be the case for high frequency, short duration pulses and small particles.

4.3.2.13 Particles Are Spherical

On the face of it, the assumption of a spherical particle is quite a restrictive one. In the case of oil crystallization, the particles formed are usually not spherical and can even form needles in the case of some fats. However, experimental measurements on such systems indicates that even in these cases, the assumption of sphericity produces good results. The experimental determination of the scattering coefficients (§4.3.9.4) overcomes any effects due to lack of sphericity, since the actual scattering properties of the system are measured, rather than theoretically predicted.

4.3.2.14 Infinite Time Irradiation

Once multiple scattering is accounted for, it is assumed that the sample is irradiated with a continuous-wave. In most experiments, however, a pulsed source is used, because it is quicker and cheaper to determine the velocity of sound this way. Scattering theory as presently formulated does not include the varying delay times from multiple scattering at different parts of a finite sample, and it is difficult to assess the effects of this. It is likely that the strongest influence on multiple scattering comes from particles which are closest together. Therefore, the correction for finite samples and short pulses is likely to be small and will leave the single scattering terms unaffected.

4.3.2.15 Pointlike Particles

Although single-particle scattering accounts for the size of particles, the multiple scattering correction does not. If the particle distribution is random, then this assumption does not give serious errors, apart from the excluded volume effects. The finite size of particles ensures that other particles are excluded from the region inside and immediately around the particle. However, because multiple scattering effects are much weaker than the single-particle effects, the overall error resulting from neglect of excluded volume is not serious.

4.3.2.16 No Overlap of Thermal and Shear Waves

In water at 1 MHz, shear waves scattered from a particle decay within 1 µm and the thermal waves decay within 0.2 µm. For many emulsions, particles may approach each other much more closely than this distance, because of diffusion, even at low concentrations and are affected by scattered waves from nearby particles. It is difficult to analyze precisely what the effect of overlap will be, and a more definitive statement must await further research.

4.3.2.17 No Interactions between Particles

Every particle is assumed to move independently of every other. If the particles are connected together in some way, say, through flocculation, their response to the sound system could have a cooperative element not accounted for within the theory as it stands. Another possible effect is the breaking and unbreaking of bonds in very weakly aggregated systems (Povey, 1996). The possible effects of weak aggregation will be considered further in Chapter 5.

4.3.2.18 Lack of Self-consistency

The incident planar wave will normally already have traveled through the dispersion, suffering scattering and rescattering. The incident wave is oscillating in an emulsion, not in the pure continuous phase. Hence, the properties of the incident wave are modified by the emulsion. This is accounted for in scattering theory as described here, by calculating average properties of the dispersion and assigning them (wavenumber and attenuation) to the incoming wave. This method cannot completely account for the actual properties of the incoming wave because it involves a circular argument. An assumption about scattering must be made to calculate the average properties required to determine the incident wave. Once the scattering has been calculated, then the incident wave properties should be recalculated, and so on, in an iterative scheme. There is an esoteric debate about this to which the very interested reader is referred (McClements *et al.*, 1990b; Gaunaurd and Wertman, 1990). In emulsions, the error due to lack of self-consistency is small. However, it may have more significant effects on the frequency-dependent terms and could be significant in particle sizing.

4.3.3 A DESCRIPTION OF WEAK SCATTERING

To understand the process of scattering, imagine a pulse of sound propagating through distilled water, considered to be a continuum with attributes of density and compressibility and other physical properties, which are independent of position and frequency.

It is important to realize that any pulse contains a number of frequencies. In particular, the pulses employed in the experiments described in this work contain frequency components many times higher in frequency than the fundamental frequency of their tone, in the case of a tone burst, or the reciprocal of its period, in the case of a square wave one period in duration. A representation of a pulse employed in the ultrasound scanner described in the previous chapter (Chapter 3) is shown in Figure 5.11 and its Fourier transform into the frequency dimension is shown in the following Figure 5.12. Nevertheless, at this stage in the argument, the frequency composition of the pulse is secondary. What is important is that the wavelength of all the components of the pulse be much greater than the diameter of the object from which they will scatter.

We confine our discussion then to the so-called "long wavelength limit," where the wavelength $\lambda \gg d$; d is the particle diameter. This also has the advantage of simplifying the analysis of the problem.

In the long wavelength limit under consideration, there are two dominant scattering phenomena. These are thermoelastic and viscoinertial scattering (Figure 4.5). To a good approximation these two scattering phenomena may be regarded as independent in the long wavelength limit, although they become intermingled outside this region (and in solid materials, but this lies outside the scope of this work). These phenomena are a result of the fact that three types of waves propagate in a pure liquid: the acoustic mode, which is detectable with a piston transducer, and a thermal mode and a shear mode, which both decay rapidly with distance.

4.3.3.1 Wave Potentials

In the following sections, the notation of vector calculus is used because it allows the equations of motion to be written in a more general way (Jeffrey, 1980), which allows the application of these equations to more complex situations than the ones

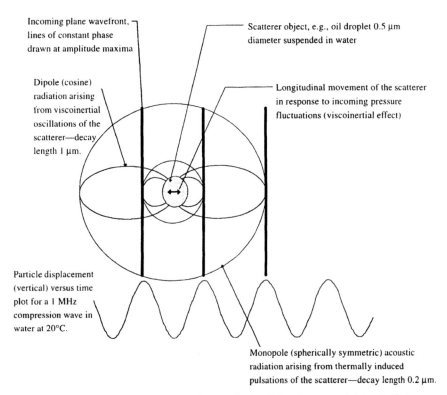

FIGURE 4.5 Ultrasound scattering from a single oil droplet suspended in distilled water. The size of the oil drop has been exaggerated relative to the wavelength of sound for clarity.

considered here. In particular, some elements of the Allegra and Hawley (1972) formulation have been adopted. Einstein notation has also been adopted, in which summation over repeated indices is implied.

The local velocity of the fluid continuum (**v**) is written in terms of wave potentials in order to simplify treatment of the boundary conditions at the scatterer/continuum boundary. The scalar quantity v, on the other hand, refers to the speed at which the sound wave propagates through the medium. Throughout this work, wherever sound or acoustic or ultrasound velocity is referred to, it is the sound speed which is meant. This is because it has now become conventional to refer to sound velocity, regardless of whether a direction is implied for the quantity. The vector **v** which refers to the local velocity of the fluid is given by

$$\mathbf{v} = -\nabla \varphi + \nabla \times \mathbf{A}, \qquad (4.7)$$

where φ is a scalar potential, and **A** is a vector potential. Equation 4.7 must be substituted into the equations of motion to obtain answers for the wavenumbers in the bulk medium. A potential is obtained from the respective field quantities (pressure, displacement, temperature, and so on) by integration with respect to distance. Conversely, the field quantity may be obtained from its potential (or potentials) by taking the appropriate gradient. The gradient operator ∇ represents differentiation with respect to space to produce a vector quantity thus:

$$\nabla = \mathbf{i}\frac{\partial}{\partial x} + \mathbf{j}\frac{\partial}{\partial y} + \mathbf{k}\frac{\partial}{\partial z}. \tag{4.8}$$

Here **i**, **j**, and **k** are the unit vectors in the x, y, and z directions, respectively.

4.3.3.2 Modes in a Pure Liquid

The acoustic mode is called a propagational mode because it can travel over considerable distances. In addition, a medium can support two other modes of acoustic wave motion: thermal and shear waves. Thermal waves are compressional acoustic waves which arise because of thermal conduction. Shear waves are due to viscous and inertial effects and are transverse waves.

Although Equations 4.9, 4.10, and 4.11 are written in different notation, the equations of motion for the passage of a sound wave through fluid media shown here are actually those given by Epstein and Carhart (1953). There are three equations—the equation of continuity (conservation of mass), the Stokes–Navier equation (conservation of momentum; this is Newton's second law written for fluids), and the energy conservation equation.

Equation for the conservation of mass:

$$\frac{\partial \rho}{\partial t} + \nabla \times \rho \mathbf{v} = 0. \tag{4.9}$$

Newton's second law, Navier–Stokes equation:

$$\rho\left(\frac{\partial \mathbf{v}}{\partial t} + (\mathbf{v} \times \nabla)\mathbf{v}\right) = -\operatorname{Div} \mathbf{P}. \tag{4.10}$$

Equation for the conservation of energy:

$$\rho\left(\frac{\partial U}{\partial t} + \mathbf{v} \times \nabla U\right) = \operatorname{Tr}\left(\mathbf{P} \times \dot{\mathbf{S}}'\right) + \tau \nabla^2 T \tag{4.11}$$

where **v** is the local velocity of the fluid continuum; **P** is the stress tensor, which includes the hydrostatic pressure p; $\dot{\mathbf{S}}'$ is the strain-rate tensor; U is the specific internal energy; τ is thermal conductivity; T is temperature; Div means divergence in vector notation; and Tr means trace in vector notation. Note that in the energy equation, the product of stress and strain rate is the rate of energy production resulting from straining the material.

To obtain the correct viscoelastic relationships, it is necessary to examine the formal relationship between stress and strain in some detail. This is nontrivial and there is not space to deal with it properly here. The equations of motion just shown apply to both liquid and solid systems, provided the correct forms for the stress and strain tensors are chosen. The situation is formally more complicated in a liquid because viscoelasticity may impart memory effects on the strain tensor.

4.3 SCATTERING THEORY

We write down the following definitions:

$$\xi = -\nabla\phi, \tag{4.12}$$

where ξ is the displacement of a volume element.

$$S'_{ij} = \frac{1}{2}\left(\frac{\partial \xi_i}{\partial x_j} + \frac{\partial \xi_j}{\partial x_i}\right)$$

$$\dot{S}'_{ij} = \left(\frac{\partial v_i}{\partial x_j} + \frac{\partial v_j}{\partial x_i}\right)$$

$$v_i = \frac{\partial \xi_i}{\partial t}$$

$$\mathbf{v} = \mathbf{i}v_x + \mathbf{j}v_y + \mathbf{k}v_z, \tag{4.13}$$

S'_{ij} is the strain tensor whose trace is

$$S = \mathrm{Tr}(S'_{ij}) = S_{ii} = \frac{\partial \xi_i}{\partial x_i} = -\nabla^2\phi, \tag{4.14}$$

and $\dot{S}'$ is the strain-rate tensor.

The stress tensor, **P**, may be related to the strain tensor in a way which connects current stresses to the deformation history (Charlier and Crowet, 1986; Tschoegl, 1989):

$$\mathbf{P} = p\delta_{ij} - \int_{-\infty}^{t}\left[\tilde{B}(t-t') - \frac{2}{3}G(t-t')\right]S(t')\delta_{ij}\, dt' - 2\int_{-\infty}^{t}G(t-t')\dot{S}'(t')\delta t'. \tag{4.15}$$

Here $\tilde{B}(t)$ is the time-dependent bulk modulus from which the reciprocal of the adiabatic compressibility (B) is absent, and $G(t)$ is the time-dependent shear modulus.

To obtain solid-like behavior, we set

$$\tilde{B}(t) = \left(\lambda + \frac{2}{3}\mu\right)h(t)$$

$$G(t) = \mu h(t), \tag{4.16}$$

where $h(t)$ is the Heavyside step function, and then we obtain the Allegra and Hawley (1972) result for a solid, the stress tensor, which is written in terms of the strain tensor (not the strain-rate tensor: as in Epstein and Carhart, 1953).

$$\begin{aligned}P_{ij} &= -\left(\lambda + \frac{2}{3}\mu\right)\mathrm{Tr}(S'_{ij})\delta_{ij} - 2\mu S_{ij}\\ &= -\left(\lambda + \frac{2}{3}\mu\right)S\delta_{ij} - 2\mu S'_{ij} + \frac{2}{3}\mu S\delta_{ij}\\ &= -\lambda S\delta_{ij} - 2\mu S'_{ij}.\end{aligned} \tag{4.17}$$

Here, λ and μ are Lamé constants; S is the trace of the strain tensor, $\mathbf{S}'$; and the traceless strain tensor S_{ij} is

$$S_{ij} = S'_{ij} - \frac{1}{3} S \delta_{ij}. \tag{4.18}$$

δ_{ij} is the Kronecker delta symbol, which has the value $\delta_{ij} = 1$ when $i = j$ and $\delta_{ij} = 0$ when $i \neq j$. Allegra and Hawley (1972) use the transformation $(G = \mu) \to (-i\omega\eta_s)$ to obtain the behavior of the liquid from the stress tensor (Equation 4.17) for the solid by using the shear viscosity η_s. This ignores the bulk viscosity (η_v) and is not general enough to apply to many viscoelastic liquids such as polymers which exhibit relaxation effects. Equation 4.15, on the other hand, is general enough to account for relaxation effects as well as more complex behaviors.

If, instead, we choose

$$\tilde{B}(t) = \eta_v \delta(t)$$
$$G(t) = \eta_s \delta(t) \tag{4.19}$$

for the case of a Newtonian liquid, where $\delta(t)$ is the Dirac function, then we obtain the stress tensor for liquid-like behavior:

$$\mathbf{P} = -p\delta_{ij} + \left(\eta_v - \frac{2}{3}\eta_s\right)\dot{S}\delta_{ij} + 2\eta_s \dot{\mathbf{S}}', \tag{4.20}$$

which is written in terms of the strain-rate tensor and agrees with the stress tensor in Epstein and Carhart (1953). This also agrees with the stress tensor for a Newtonian fluid given by Batchelor (1967) and Morse and Ingard (1968).

Ignoring relaxation effects (the $\tilde{B}(t - t')$ in Equation 4.15) which may be important in polymers, the principal stress is:

$$p = \frac{1}{3}\mathrm{Tr}(\mathbf{P}) = -\left(\lambda + \frac{2}{3}\mu\right)S = -BS. \tag{4.21}$$

The principal stress p in Equation 4.21 is simply the hydrostatic pressure, and the equation relates pressure to the fractional change in volume, S, through the bulk modulus, B. This equation also relates the Lamé constants to the bulk modulus in the case of a solid.

The preceding equations of motion, together with the stress tensor in Equation 4.20 and the thermodynamic relations in the medium, give rise to the following *dispersion relations* (a relationship between the wavenumber and the frequency):

$$\frac{\frac{1}{k^2}}{\frac{1}{k_t^2}} = \frac{v^2}{2\omega^2}\left\{1 - i(e_k + \gamma f_k) \pm \left[1 - 2ie_k - 2if_k(\gamma - 2) - (e_k - \gamma f_k)^2\right]^{1/2}\right\}, \tag{4.22}$$

where v is the speed of sound; $\gamma = C_p/C_v$ is the ratio of the specific heats; and e_k and f_k are given by

4.3 SCATTERING THEORY

$$e_k = \frac{4}{3}\frac{(\eta_s + B/2)\omega}{\rho v^2}, \quad f_k = \frac{\sigma\omega}{v^2} \qquad (4.23)$$

$$k_s = (1+i)\sqrt{\frac{\omega\rho}{2\eta_s}}, \qquad (4.24)$$

where η_s/ρ is the kinematic viscosity, B is the adiabatic bulk modulus if the material is solid, and B is the coefficient of compressional viscosity (η_v) if the material is liquid (Epstein and Carhart, 1953). A very good approximation for the thermal mode is

$$k_t = (1+i)\sqrt{\frac{\omega}{2\sigma}}, \qquad (4.25)$$

where $\sigma (= \tau/\rho C_p)$ is called the *thermometric conductivity* or *thermal diffusivity*. The imaginary part of both Equations 4.24 and 4.25 is equal to the real part. This means that both the thermal and shear modes are dissipative and will be strongly attenuated over a distance of only a few wavelengths. The thermal mode is the means by which energy is lost through thermal conduction and the shear mode by viscous dissipation. Because these are dissipative, not propagational modes, they are only present in the region around the interface where they are produced and do not contribute to the acoustic field over a long range. Their importance to us is that they are produced at interfaces where the properties of the medium change and are an important part of the scattering process.

4.3.3.3 Thermoelastic Scattering

Thermoelastic scattering affects the compressibility of the system. Associated with the pressure wave that comprises the ultrasound pulse is a temperature wave (Figure 4.2) which under adiabatic conditions is in step with the pressure wave. Examination of Table 2.1 indicates that this temperature wave will have an amplitude of a few millikelvins. It is tempting to assume that consequently it can be neglected and this has been the course pursued, incorrectly, by many. However, this temperature wave can have profound effects on the propagation of the pulse once it meets a scatterer. The specific heat of an oil droplet (1980 J kg^{-1} K^{-1}) for sunflower oil (McClements and Povey, 1989) is about half that for water (4182 J kg^{-1} K^{-1}). So although in water the temperature fluctuation may be on the order of a few millikelvins, in the bulk of the oil droplet the same pressure fluctuation will produce temperature fluctuations of twice that, all other things being equal. At the boundary between the oil drop and the water a discontinuity in the temperature would exist if heat could not flow, and the adiabatic condition will therefore be violated at the boundary. Heat therefore flows from and to the surface of the oil drop, depending on whether it is in compression or rarefaction and on the sign of the coefficient of volume expansivity, which is negative in water below 4°C and positive in most other liquids. This heat flow will heat and cool the boundary layer around the oil drop with an associated expansion and contraction of the boundary layer. Thus, the drop becomes a secondary source of sound, with spherically symmetric pulsation out of step with the original sound pressure wave. (See Figure 4.5). In addition, a thermal wave, which helps maintain the continuity of temperature across the boundary, will flow away from the drop and die away. The distance over which the amplitude of the thermal wave decays to $1/e$ of its maximum amplitude is called the *skin depth* or *decay length* of the wave. This distance is given by the relationship

$$\delta_t = \sqrt{\frac{2\sigma}{\omega}}, \tag{4.26}$$

where δ_t is the *thermal skin depth* or *thermal wave decay length*. All variables are properties of the continuous phase.

4.3.3.4 Viscoinertial Scattering

The source of viscoinertial scattering lies in the difference in density, and hence inertia, between the scatterer and its surroundings. In the long wavelength region viscoinertial scattering may be viewed as modifying the density of the mixture. The oscillating forces associated with the sound pulse result in motion of the scatterer relative to its surroundings. This motion increases with increasing density difference between the scatterer and the fluid, but the fluid reacts back on the scatterer through the displacement of excluded volume and hence fluid motion occurs around the scatterer. The larger the viscosity of the fluid, the more resistance to the motion will occur. Large viscosity can completely cancel out the effects of inertial difference, preventing relative motion from occurring. The relative motion is itself a source of sound as the material obstructing the scatterer motion is compressed and the material in its wake rarefied. The sound field produced has a cosine dependence on angle and is opposite in phase in the forward and backward directions (Figure 4.5). In addition, a shear field will propagate away from the drop, but because fluid cannot support shear to any significant extent, it will die away over a short distance relative to the size of the drop.

$$\delta_s = \sqrt{\frac{2\nu}{\omega}}, \tag{4.27}$$

where δ_s is the shear wave decay length, and $\nu\, (= \eta_s/\rho)$ is the kinematic viscosity of the continuous phase.

4.3.3.5 Scattered Waves Combine within the Transducer

At a long distance away from the scatterer, relative to its size, the wavefronts of the sound waves scattered by the drop will appear to all intents and purposes plane. The conventional ultrasound receiver, manufactured from piezoelectric material, is sensitive to the phase of the signal and integrates the pressure across its face. What is detected is the sum of the original pulse and the scattered wave. In general this signal may have both its amplitude and its phase changed by the presence of the

4.3 SCATTERING THEORY

scattered signal. The change in phase is normally interpreted as a change in the velocity of sound in the medium, since the incoming pulse may have its position in time advanced or delayed by the scattering.

4.3.4 PLANE WAVE INCIDENT ON A SINGLE PARTICLE

4.3.4.1 Introduction

The treatment of scattering in this work is based on analyses of the scattering of a plane wave incident on a single particle, carried out by Epstein and Carhart (1953) and Allegra and Hawley (1972). The two works complement each other, and the Allegra and Hawley paper contains a number of typographical errors. The results in this section are an amalgamation of the most detailed results from both papers.

The incident mode is a propagational mode, and at the surface of the particle other modes are produced, both inside and outside the particle. The acoustic field observed a long way from the particle is the sum of the incident wave and the outgoing propagational wave produced at the particle surface (Figures 4.3 and 4.5). As explained earlier, a change of sound velocity is apparent because the addition of the incident and scattered waves produces a phase shift at the receiver transducer. Attenuation is a result of the energy dissipated in the thermal and shear modes both inside and outside the particle. An additional source of attenuation arises because acoustic energy is scattered out of the forward direction and is radiated in all directions.

Important Note: The attenuation computed from scattering theory is in addition to any attenuation inherent in the liquid making up the continuous phase. The attenuation inherent in the continuous phase is assumed to be small (§4.3.2.2).

4.3.4.2 Spherical Harmonics

Partial wave analysis, in which each wave potential is described by a sum of waves of differing angular (spherical, dumbbell, etc.) and radial dependencies, permits the various components of the wave to be independently matched at the particle boundary. Omitting the time variation, the incident plane wave can be written in terms of all the possible angular (Legendre polynomials, $P_n(\cos(\theta))$) and spherical (Bessel functions, $j_n(kR)$) variations out of which it can be constructed, in a somewhat similar way to the way in which a wave can be written as the sum of its Fourier components. Note that the spherical coordinates used here are (R, θ, ϕ), whereas r is the radius of the scatterer. The spherical coordinate system has its origin at the center of the particle, which is assumed to be spherical in order to simplify the problem. A single frequency is considered and the time variation is omitted for clarity, so that the incident plane wave may be written as

$$\varphi_0 = e^{ikz} = \sum_{n=0}^{\infty} i^n (2n+1) j_n(kR) P_n(\cos\theta). \quad (4.28)$$

Here the direction of propagation is the z direction, n is the order of the polynomial, and θ is the angle with respect to the z direction. $i = \sqrt{-1}$. The plane incident wave, φ_0 is represented in this complicated way so that each of the components of the incident wave can be matched with the respective components of the scattered wave at the spherically symmetrical particle boundary.

All workers in this field omit a constant factor in the wave potential φ which represents the actual amplitude of the sound wave. The acoustic intensity in a plane wave is equal to the sound speed multiplied by the acoustic energy density (Pierce, 1981a, p. 39). This energy density is twice the kinetic energy per volume due to the sound field. Equation 4.8 and Equation 4.28 can then be used to show that this amplitude factor is

$$\sqrt{\frac{I}{\rho k^2 v}}, \qquad (4.29)$$

where I is the acoustic intensity, which is the acoustic power flowing through unit area normal to the propagation direction. All parameters which are proportional to the incident scalar potential φ_0 must be multiplied by this factor.

4.3.4.3 Boundary Conditions

The scattered waves produced at the particle boundary can be written in a similar form to Equation 4.28, but this time the coefficients are to be determined by application of the appropriate boundary conditions. Because we consider only a plane incident wave, the vector potential of the shear wave reduces by symmetry to a single component in the azimuthal direction, written $A_\psi = A$. Again the amplitude factor (Equation 4.29) is omitted. The wave potentials for the scattered waves at the particle surface are

$$\varphi_R = \sum_{n=0}^{\infty} i^n (2n+1) A_n h_n(kR) P_n(\cos\theta) = \sum_{n=0}^{\infty} \varphi_{Rn}$$

$$\varphi_t = \sum_{n=0}^{\infty} i^n (2n+1) B_n h_n(k_t R) P_n(\cos\theta)$$

$$A = \sum_{n=1}^{\infty} i^n (2n+1) C_n h_n(k_s R) P_n^1(\cos\theta)$$

$$\varphi' = \sum_{n=0}^{\infty} i^n (2n+1) A_n' j_n(k'R) P_n(\cos\theta)$$

$$\varphi_t' = \sum_{n=0}^{\infty} i^n (2n+1) B_n' j_n(k_t'R) P_n(\cos\theta)$$

$$A' = \sum_{n=1}^{\infty} i^n (2n+1) C_n' j_n(k_s'R) P_n^1(\cos\theta), \qquad (4.30)$$

4.3 SCATTERING THEORY

where j_n and h_n are the spherical Bessel and Hankel functions, and P_n and P'_n are the Legendre and associated Legendre polynomials respectively. Primed symbols refer to the interior of the particle. Here the subscript R refers to the scattered propagational mode and t to the thermal mode. The zero-order scattered acoustic mode (φ_{R0}) represents the result of the particle pulsation, arising from the temperature difference and the compressibility difference between the particle and its suspending fluid. It is spherically symmetric and is sometimes called *monopolar*. The first-order mode (φ_{R1}) represents the movement of the particle backwards and forwards in the direction of the incident wave (Figure 4.5). This is called dipolar radiation. In the long wavelength limit, the series in the ϕ's can usually be truncated at the first-order mode, and for this reason partial wave analysis is ideal for the analysis of acoustic propagation in this limit. At higher frequencies, more and more terms in the series must be taken into account. All that is necessary in practice, therefore, is to determine the two coefficients A_0 and A_1. These are called the single-particle scattering coefficients because they are usually all that is needed from Equation 4.30 to determine the velocity and attenuation of the wave at large distances from the particle. Ten boundary conditions must be applied to determine these two coefficients. The boundary conditions are continuity of stress normal to the surface, velocity, temperature, and heat flux at the surface of the particle. Each mode apart from the zeroth mode has six boundary conditions because there are six unknowns. The zeroth mode has four conditions because no shear waves arise in this mode. So a total of 10 conditions are required to determine the zeroth- and first-order modes. The boundary conditions for fluid particles were given by Epstein and Carhart (1953) and for solid particles by Allegra and Hawley (1972). The conversion from one to the other is interesting. The fluid viscosity is replaced by the shear modulus of the solid, as follows

$$\eta_s = \frac{G}{(-i\omega)}, \qquad (4.31)$$

and all terms for the solid material are multiplied by a factor of $-i\omega$. Details may be found in Allegra and Hawley (1972). G is the shear modulus for the solid material. There is no doubt that this approach leaves something to be desired. For example, it omits to account for the bulk viscosity. A more general approach is required, involving the stress tensor and the inclusion of a constitutive equation for a viscoelastic fluid which relates the current stresses to the deformation history (Equation 4.15).

To obtain the boundary conditions at the particle surface, physical quantities such as temperature and stress must be expressed in terms of wave potentials. The fluid velocity is derived from Equation 4.7 to Equation 4.30. The temperature potential is derived from the energy equation and thermodynamic relations and is found to be scalar. The shear mode does not contribute to any temperature change. The scalar potential is related to the temperature through

$$\Delta T = \Gamma \varphi, \qquad (4.32)$$

where ΔT is the temperature change associated with the acoustic wave, and

$$\Gamma = \frac{-ik^2(\gamma-1)}{\omega\beta\left(1+\frac{i\gamma\sigma k^2}{\omega}\right)}, \qquad (4.33)$$

where β is the volume coefficient of thermal expansivity and σ is the thermometric conductivity. k in Equation 4.33 is either k for the propagational acoustic mode or k_T for the thermal mode.

The stress component P_{rr} is made up of two terms: the pressure and the viscous stress. All other terms have been defined earlier in this chapter. For the scalar modes, which contribute to the hydrostatic pressure, these terms combine to give

$$P_{rr} = (i\omega\rho - 2\eta_s k^2)\varphi - 2\eta_s \frac{\partial^2\varphi}{\partial R^2}. \qquad (4.34)$$

The shear mode only includes viscous stress terms:

$$P_{rr} = \frac{2\eta_s}{\sin\theta}\frac{\partial}{\partial\theta}\left[\sin\theta\left(-\frac{A}{R^2}+\frac{1}{R}\frac{\partial A}{\partial R}\right)\right]. \qquad (4.35)$$

The other stress component, $P_{r\theta}$, is given by

$$P_{r\theta} = 2\eta_s \frac{\partial}{\partial\theta}\left(\frac{\varphi}{R^2} - \frac{1}{R}\frac{\partial\varphi}{\partial R}\right) \qquad (4.36)$$

for the compressional modes and

$$P_{r\theta} = \eta_s\left[\left(\frac{2A}{R^2}-\frac{\partial^2 A}{\partial R^2}\right)+\frac{1}{R^2}\frac{\partial}{\partial\theta}\left(\frac{1}{\sin\theta}\frac{\partial}{\partial\theta}(A\sin\theta)\right)\right] \qquad (4.37)$$

for the transverse (shear) mode.

The following set of equations define the boundary conditions for fluid particles in a fluid medium in the order radial velocity, temperature, heat flux, P_{rr} stress component, tangential velocity and $P_{r\theta}$ stress component. These equations ensure that the corresponding *physical* quantities are continuous across the scatterer boundary.

(a) $aj'_n(a) + A_n a h'_n(a) + B_n b h'_n(b) - C_n n(n+1) h_n(c)$

$= A'_n a' j'_n(a') + B'_n b' j'_n(b') - C'_n(n+1) j_n(c')$

(b) $G_p j_n(a) + G_p A_n h_n(a) + G_t B_n h_n(b) = G'_p A'_n j_n(a') + G'_t B'_n j_n(b')$

4.3 SCATTERING THEORY

(c) $\tau\{G_p aj'_n(a) + G_p A_n ah'_n(a) + G_t B_n bh'_n(b)\} = \tau'\{G'_p A'_n a'j'_n(a') + G'_t B'_n b'j'_n(b')\}$

(d) $\left[(i\omega\rho - 2\eta_s k^2)j_n(a) - 2\eta_s k^2 j''_n(a)\right] + A_n\left[(i\omega\rho - 2\eta_s k^2)h_n(a) - 2\eta_s k^2 h''_n(a)\right]$

$+ B_n\left[(i\omega\rho - 2\eta_s k_t^2)h_n(b) - 2\eta_s k_t^2 h''_n(b)\right] + 2n(n+1)\eta_s C_n \dfrac{\left[ch'_n(c) - h_n(c)\right]}{r^2}$

$= A'_n\left[(i\omega\rho' - 2\eta'_s k'^2)j_n(a') - 2\eta'_s k'^2 j''_n(a')\right]$

$+ B'_n\left[(i\omega\rho' - 2\eta'_s k_t'^2)j_n(b') - 2\eta'_s k_t'^2 j''_n(b')\right]$

$+ 2n(n+1)\eta'_s C'_n \dfrac{\left[c'j'_n(c') - j_n(c')\right]}{r^2}$

(e) $j_n(a) + A_n h_n(a) + B_n h_n(b) - C_n\left[h_n(c) + ch'_n(c)\right]$

$= A'_n j_n(a') + B'_n j_n(b') - C'_n\left[j_n(c') + c'j'_n(c')\right]$

(f) $\eta_s\left[aj'_n(a) - j_n(a)\right] + \eta_s A_n\left[ah'_n(a) - h_n(a)\right] + \eta_s B_n\left[bh'_n(b) - h_n(b)\right]$

$-\dfrac{1}{2}\eta_s C_n\left[c^2 h''_n(c) + (n^2 + n - 2)h_n(c)\right]$

$= \eta'_s A'_n\left[a'j'_n(a') - j_n(a')\right] + \eta'_s B'_n\left[b'j'_n(b') - j_n(b')\right]$

$-\dfrac{1}{2}\eta'_s C'_n\left[c'^2 j''_n(c') + (n^2 + n - 2)j_n(c')\right]$, (4.38)

where G_p and G_t are the thermal factors calculated from Equation 4.33 for the propagational and thermal modes, respectively. A primed function represents a spatial derivative. Otherwise, prime denotes the dispersed phase inside the particle.

The solution of this set of equations for each mode (denoted by n) produces a value for the single-particle scattering coefficient, A_n, of that mode. The fifth and sixth of these equations (for the tangential velocity and the stress $P_{r\theta}$) are invalid for the zero-order mode, because the angular function factor which was cancelled throughout is zero in this case. No shear waves are produced in the zero-order mode. Although analytical results can be obtained in special cases, the general solution for the scattering coefficients must be calculated numerically.

While these equations are daunting, they have been solved by many workers and have been shown to give accurate results. FORTRAN computer programs exist, including that of Pinfield (1996) which can be used to solve the equations to any order. However, explicit solutions of these equations exist under limiting conditions. These explicit solutions were first given by Allegra and Hawley (1972) and will be given later, since they are amenable to solution by computer programs such as MathCad.

4.3.5 SCATTERING BY MANY PARTICLES

4.3.5.1 Introduction

The result for the attenuation coefficient was obtained by Epstein and Carhart (1953) and Allegra and Hawley (1972) by calculating the energy in the propagational mode at large distances where the thermal and shear modes have disappeared. The total energy lost was simply assumed to be proportional to the number of particles. The result for the scattering contribution, α_s, to the total attenuation is

$$\alpha_s = -\frac{3\phi}{2k_1^2 r^3} \sum_{n=0}^{\infty} (2n+1) \operatorname{Re} A_n . \tag{4.39}$$

This result (α_s) is the attenuation contribution from the scatterers to the total attenuation in the system. It is worth noting that this result is the result for a single-particle, multiplied by the number of particles in unit volume to give the attenuation per unit distance

$$\alpha_{total} = \alpha_s + (1-\phi)\alpha_1 + \phi\alpha_2, \tag{4.40}$$

where α_{total} is the attenuation measured in the system, α_1 is the attenuation measured in the continuous phase, and α_2 is the attenuation measured in the pure dispersed phase.

4.3.5.2 Multiple Scattering Theories

The single-particle scattering approaches of Epstein and Carhart (1953) and Allegra and Hawley (1972) are limited to dilute dispersions or very weak scattering. Under these conditions, the proportion of the acoustic field incident on a particle which has been scattered by other particles is very small indeed. As the concentration of scatterers increases, the proportion of the incident field which is due to all other scatterers may become significant. Multiple scattering theory is an appropriate tool in this case. Inspection of Figure 4.4 will indicate that the computation of the phase and amplitude relationships between each stage of multiple scattering (first, second, third scattering and so on) is likely to be nontrivial. This is indeed so.

An early attempt at a multiple scattering model was tried by Urick and Ament (1949). They used a "thin slab approximation" to obtain the effective wavenumber of an emulsion in terms of the single-particle scattering amplitudes. The forward and backward scattered wave amplitudes were calculated from a thin slice of dispersion which was much smaller than one wavelength. The scattering coefficients so calculated were compared with the reflected and transmitted waves which would be obtained from a homogeneous slice of fluid with a given wavenumber. In this way the apparent wavenumber of the dispersion was determined. The scattering calculation assumed that each particle in the slice experienced the same acoustic field, neglecting transverse multiple scattering. Their result therefore omits some terms.

Two theories which are widely used for the analysis of ultrasound measurements are by Lloyd (1967a, b), Lloyd and Berry (1967), Ziman (1966), Waterman and Truell (1961), and Fikioris and Waterman (1964). Waterman and Truell (1961) determined the wavenumber in a dispersion by using the hierarchy method and ensemble averaging to obtain the acoustic field (Foldy, 1945; Lax, 1951, 1952). Ensemble averaging takes an average over all possible configurations of particle positions. The hierarchy is formed from the relationship between the ensemble average with n particles in fixed positions to the ensemble average with $n+1$ particles fixed. The observable field is the ensemble average with no particle positions fixed. Waterman and Truell's first attempt (1961) agreed with Epstein and Carhart (1953), with Allegra and Hawley (1972), and with Urick and Ament (1949). A later attempt by Fikioris and Waterman (1964) corrected a previous error and gave

$$\left(\frac{K^2}{k_1^2}\right) = 1 - \frac{3i\phi}{k_1^3 r^3}(A_0 + 3A_1) - \frac{27\phi^2}{k_1^6 r^6}(A_0 A_1 + 2A_1^2). \tag{4.41}$$

The work of Lloyd and Berry approached the problem from the point of view of the density of energy states in a medium, and the calculation of the acoustic wavenumber is produced almost as a side issue. It has been found to produce reliable results in many systems and has been widely adopted as the correct form of the multiple scattering theory result. The result presented here is terminated after the second-order scattering terms, since these are usually sufficient. The result of Lloyd and Berry for the wavenumber of a dispersion (K) is

$$\left(\frac{K^2}{k_1^2}\right) = 1 - \frac{3i\phi}{k_1^3 r^3}\left(\sum_{n=0}^{\infty}(2n+1)A_n\right)$$
$$- \frac{27\phi^2}{k_1^6 r^6}\left(A_0 A_1 + \frac{10}{3} A_0 A_2 + 2A_1^2 + 11A_1 A_2 + \frac{230}{21} A_2^2\right). \tag{4.42}$$

This result agrees with that obtained by Fikioris and Waterman (1964) to the first-order scattering mode at which it is normally truncated. Higher orders have not been compared. Here ϕ is the volume fraction occupied by scatterers and A_0 and A_1 are the zero- and first-order scattering coefficients.

It is recommended that anyone wishing to use higher orders than the second, should use the Fikioris and Waterman approach to the calculation because it is easier to understand than the Lloyd and Berry approach.

4.3.6 NUMERICAL CALCULATIONS USING SCATTERING THEORY

Numerical calculation is restricted to the low-frequency region by the behavior of the spherical Bessel functions. The periodicity of the spherical Bessel functions makes it impossible to obtain an accurate value, as the imaginary part of the argument increases for thermal and shear waves. Anson and Chivers (1993) suggest how such calculations may be extended beyond the low-frequency region. However, the majority of applications considered in this work satisfy the long wavelength limit, so special techniques are not considered here. For the first-order scattering mode, the terms in the last boundary condition of Equation 4.38 for $P_{r\phi}$ should be modified using the relationship

$$aj_1''(a) - j_1(a) = -aj_2(a) \tag{4.43}$$

for both the Bessel and the Hankel functions. For small arguments, the difference on the left-hand side of this equation is very small, and so is more accurately calculated from the second-order function.

4.3.6.1 Particle Size Distribution and Change in Phase

In the case of a distribution of particle sizes in a dispersion, the terms

$$\frac{\phi A_n}{r^3} \tag{4.44}$$

in Equation 4.42, should be replaced by

$$\sum_i \frac{\phi_i A_{ni}}{r_i^3}, \tag{4.45}$$

where the sum is over the different scatterer size classes. The single-particle scattering coefficient must be determined for each size of particle and for each type of particle (solid or liquid, for instance) separately. The combined scattering coefficient can then be calculated, given the particle size distribution and the proportion of particles which are solid.

4.3.7 THE RESULTS OF SCATTERING THEORY

McClements and Povey (1989) first applied scattering theory to a practical emulsion, sunflower oil-in-water, which forms the basis for many margarine formulations. Pinfield (1996) extended this work to include dispersions containing mixtures of phases (liquid/solid, etc.), reviewed modern acoustic scattering theory, and developed simplified approaches to the practical analysis of experimental scattering data.

In Figures 4.6 and 4.7, the ultrasound velocity and attenuation for a monodisperse sunflower oil-in-water system represented by the data in Table 4.2 are given, plotted against a scaling factor, $f^{0.5}r$, which combines the effects of particle size and frequency on velocity and attenuation per wavelength ($\alpha\lambda$) in a convenient way.

4.3 SCATTERING THEORY

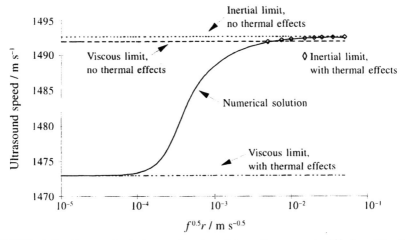

FIGURE 4.6 Ultrasound velocity as a function of the scaling parameter $f^{0.5}r$ for a 20 vol% sunflower oil-in-water emulsion at 30°C (Pinfield, 1996).

There are three regions within the long wavelength limit which applies to Figures 4.6 and 4.7. At very long wavelengths ($\lambda \to \infty$, $r \to 0$, $f \to 0$, ϕ constant) the velocity is independent of frequency and particle size. This region is convenient for phase volume determination because it is simpler if particle-size and frequency-dependent effects can be ignored. However, this region is affected by the thermal properties of the system. In this region, the attenuation scales with the square of the frequency. At shorter wavelengths, but still within the long wavelength limit, the velocity again varies very little with frequency and the attenuation scales with the

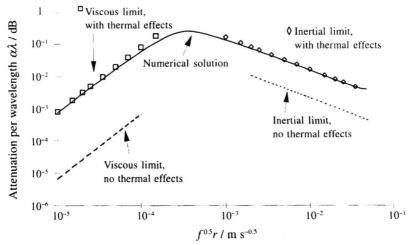

FIGURE 4.7 Ultrasound attenuation per wavelength as a function of the scaling parameter $f^{0.5}r$ for a 20 vol% sunflower oil-in-water emulsion at 30°C (Pinfield, 1996).

TABLE 4.2 Physical properties for sunflower oil and pure water at 30°C[a]

	Water	Sunflower oil
Ultrasound velocity / m s^{-1}	1509.127	1437.86
Density / kg m^{-3}	995.65	912.9
Shear viscosity / Pa s	0.0007973	0.054
Thermal conductivity / J m^{-1} s^{-1} K^{-1}	0.6032	0.170
Specific heat capacity (at constant pressure) / J kg^{-1} K^{-1}	4178.2	1980.0
Thermal expansivity / K^{-1}	0.000301	0.0007266

[a] The data for water were taken from Kaye and Laby (1986). The properties of sunflower oil were given by McClements (1988).

square root of the frequency. This region merges into the geometric limit and can even disappear entirely under some circumstances. The intermediate region shows a complicated dependence on frequency and particle size, and it is in this region that ultrasonic particle sizing may be applied, since the velocity of sound is sensitive to size. One important consequence of the existence of a very long wavelength limit is that it offers the possibility of obtaining simplified solutions to the scattering theory results. Allegra and Hawley (1972) showed that it was possible to obtain explicit solutions to the scattering coefficients under some circumstances, and these can be used to obtain approximations in the intermediate region.

4.3.8 SIMPLIFIED SCATTERING COEFFICIENTS

If we define a new scattering coefficient a_n in terms of the original coefficients A_n thus,

$$a_n = -\frac{3iA_n}{k^3 r^3}, \tag{4.46}$$

the coefficients a_n are called the modified scattering coefficients.

Then Equation 4.41 can be further simplified:

$$K = k_1 \left[1 + \frac{\phi}{2}(a_0 + 3a_1) - \frac{\phi^2}{8}(a_0^2 - 6a_0 a_1 - 15a_1^2) \right]. \tag{4.47}$$

Consequently, the equations for the velocity and attenuation, accurate to the first-order scattering coefficient, may be written

$$\frac{1}{v^2} = \frac{1}{v_1^2}\left[(1+\phi\,\mathrm{Re}\,a_0)(1+3\phi\,\mathrm{Re}\,a_1)+6\phi^2(\mathrm{Re}\,a_1)^2\right] \tag{4.48}$$

4.3 SCATTERING THEORY

and

$$\alpha_s = \frac{\omega}{v_1}\left[\frac{\phi}{2}\text{Im}(a_0+3a_1)-\frac{\phi^2}{8}\text{Im}(a_0^2-6a_0a_1-15a_1^2)\right]$$
$$+\alpha_1\left[1+\frac{\phi}{2}\text{Re}(a_0+3a_1)-\frac{\phi^2}{8}\text{Re}(a_0^2-6a_0a_1-15a_1^2)\right]. \quad (4.49)$$

In Equation 4.48, the facts that the attenuation in the continuous phase is small and that the imaginary part of the modified scattering coefficients is much smaller than the real part have been used to obtain further simplification. Study of Equations 4.48 and 4.49 indicates that the velocity is given by the real part of combinations of the modified scattering coefficients and the attenuation by the imaginary part of combinations of the modified scattering coefficients. This is because Equation 4.46 exchanges the real and imaginary parts between A_n and a_n.

The conditions for the limiting analytical cases of the modified single-particle scattering coefficients for fluid particles in a fluid medium are given in Table 4.3. The zero-order mode represents the pulsation of the particle generating spherically symmetric (monopole) scattering (see Figure 4.5) and the corresponding unmodified scattering coefficient is

$$A_0 = \frac{ik^3r^3}{3}\left(\frac{\rho k'^2}{\rho' k^2}-1\right)+ikr\frac{G_p}{G_t}\left(1-\frac{\tau G_t}{\tau' G_t'}\right)^2 \times \frac{1}{F}, \quad (4.50)$$

where

$$F = \frac{h_0(b)}{bh_1(b)}-\frac{\tau}{\tau'}\frac{j_0(b')}{b'j_1(b')} \quad (4.51)$$

TABLE 4.3 Conditions for the Limiting Values of the Modified Single-Particle Scattering Coefficients for Fluid Particles in a Fluid Medium[a]

Condition	Description
$kr \ll 1$, $k'r \ll 1$	Propagational wavelength much larger than particle radius (Rayleigh limit)
$\left\|\frac{k^2}{k_t^2}\right\| \ll 1$, $\left\|\frac{k'^2}{k_t'^2}\right\| \ll 1$	Thermal wavelengths much smaller than the propagational wavelengths
$\left\|\frac{k^2}{k_s^2}\right\| \ll 1$, $\left\|\frac{k'^2}{k_s'^2}\right\| \ll 1$	Shear wavelengths much smaller than the propagational wavelengths
$\alpha^2 \ll \frac{\omega^2}{v^2}$, $\alpha'^2 \ll \frac{\omega^2}{v'^2}$	Attenuation in bulk phases negligible

[a] Pinfield (1996).

(see Epstein and Carhart, 1953; Allegra and Hawley, 1972). The zero-order coefficient can be written in two parts as

$$A_0 = A_{01} + A_{02}. \tag{4.52}$$

The first part of the zero-order scattering coefficient is unaffected by thermal effects and depends on the compressibility difference between the two phases. Expressions for the two parts of the zero-order scattering coefficient are given in Table 4.4.

It should be noted that the use of the limiting expressions in Table 4.4 does not simply reproduce the results of Allegra and Hawley (1972) but gives a more accurate approximation because Allegra and Hawley used the single scattering form of the attenuation (Equation 4.39). Moreover, a number of typographical errors in Allegra and Hawley necessitate great care in reproducing their work.

The second term in the zero-order scattering coefficient incorporates thermal effects. For fluids this is a result of the coupling of temperature and pressure in the

TABLE 4.4 Limiting Analytical Expressions for the Zero-Order Single-Particle Scattering Coefficient for Fluid Particles, Satisfying the Criteria in Table 4.3[a]

Scattering coefficient	Condition	Description
$A_{01} = \dfrac{ik^3 r^3}{3} \times \left\{ \dfrac{\Delta \kappa}{\kappa} + 2i \dfrac{\kappa'}{\kappa} \Delta\left(\dfrac{v\alpha}{\omega}\right) \right\}$		
$A_{02} = \dfrac{ik^3 r^3 (\gamma - 1)}{3} \times \dfrac{\rho' C_p'}{\rho C_p} \dfrac{\left[\Delta\left(\dfrac{\beta}{\rho C_p}\right)\right]^2}{\left(\dfrac{\beta}{\rho C_p}\right)} \times \left\{ 1 + \dfrac{i\omega r^2 \rho' C_p'}{3\tau}\left(1 + \dfrac{\tau}{5\tau'}\right) \right\}$	$\|k_t r\| \ll 1,$ $\|k_t' r\| \ll 1$	Thermal wavelength much larger than particle radius
$A_{02} = \dfrac{ik^3 r^3}{3} \times \dfrac{3(\gamma - 1)}{r\sqrt{2\omega}} \dfrac{\left[\Delta\left(\dfrac{\beta}{\rho C_p}\right)\right]^2}{\left(\dfrac{\beta}{\rho C_p}\right)} \times \sqrt{\dfrac{\tau\tau'\rho' C_p'}{\rho C_p}} \times \dfrac{(1+i)}{\left(\sqrt{\tau'\rho' C_p'} + \sqrt{\tau\rho C_p}\right)}$	$\|k_t r\| \gg 1,$ $\|k_t' r\| \gg 1$	Thermal wavelength much shorter than particle radius

[a] Pinfield (1996).

4.3 SCATTERING THEORY

different materials. For a given pressure, the temperature deviation generated is proportional to the parameter

$$\frac{\beta}{\rho C_p}. \tag{4.53}$$

Thermal acoustic modes are generated in order to maintain continuity of temperature at the particle surface, given the different temperature changes in the differing materials making up particle and continuum, for the same pressure field. The different heat flow properties inside and outside the particle also contribute. Note from Table 4.4 that there are two different limiting forms for the thermal contribution to the zero-order scattering coefficients, thermal wavelength much greater than particle size and thermal wavelength much shorter than particle size. Since the thermal wavelength is of the order of 0.2 μm at 1 MHz in water at 20°C, either of these conditions is likely to be encountered.

The limiting analytical expressions for the first-order scattering coefficients are given in Table 4.5. The first-order coefficients generate a dipole (cosine or dumbbell) distribution of intensity with the greatest intensity in the forward and backward direction (along the z axis, see Figure 4.5). This is a result of viscoinertial scattering (McClements and Povey, 1989). There are two important limits in the case of viscoinertial scattering.

When the shear wavelength is much larger than the particle size, the viscous drag of the surrounding fluid exceeds the inertial forces on the particle. The particle moves in phase with the fluid medium in this case. It is possible for the inertial effects to reimpose themselves in this region but very large density differences are required for this to occur. This is an unusual situation in most dispersions because

TABLE 4.5 Limiting Analytical Expressions for the First Order Unmodified Scattering Coefficients for Fluid Particles in a Fluid Medium[a]

Scattering coefficient	Condition	Description
$A_1 = \dfrac{ik^3 r^3}{9} \times \dfrac{\Delta\rho}{\rho}\left[1 + \dfrac{2i}{9} \times \dfrac{\omega\rho r^2}{\eta_s} \times \dfrac{\Delta\rho}{\rho}\right]$	$\|k_s r\| \ll 1$, $\|k'_s r\| \ll 1$ $\eta'_s \gg \eta_s$	<u>Viscous limit/low frequency limit</u> • Shear wavelength much larger than particle radius • Particle very viscous • Viscous effects dominate over inertial effect
$A_1 = \dfrac{ik^3 r^3}{3} \times \dfrac{(\rho' - \rho)}{(2\rho' + \rho)} \times$ $\left[1 + \dfrac{3i}{r}\sqrt{\dfrac{2\eta_s}{\omega\rho}} \times \dfrac{(\rho' - \rho)}{(2\rho' + \rho)}\right]$	$\|k_s r\| \gg 1$, $\|k'_s r\| \gg 1$ $\eta'_s \gg \eta_s$	<u>Inertial limit</u> • Shear wavelength much shorter than particle radius • Particle very viscous • Inertial effects dominate over viscous effects

[a]Pinfield (1996).

the equalization of density differences is one of the simplest ways of achieving an emulsion which is stable under gravity. In this limit, the scattering coefficient depends on the density difference between particle and fluid and is not governed by the response of the material inside the particle to the sound wave.

If the shear mode wavelength is much less than the particle size, the effect of inertia exceeds that of the viscous drag and the particle moves relative to its surroundings. The fluid then appears to the sound wave to have acquired an additional inertia due to the presence of the particle, and the scattering coefficient is shown in Table 4.5. This may be termed the *inertial limit* here, the contribution of the inertial effect to the velocity is independent of frequency and the attenuation varies as the square root of frequency.

4.3.9 WORKING EQUATIONS

4.3.9.1 The Urick Equation

The simplified scattering coefficients presented in Tables 4.4 and 4.5 above may be inserted into the equations for the velocity (Equation 4.48) and attenuation (Equation 4.49) to obtain simplified and explicit expressions under the relevant limiting conditions. Following this method, Pinfield *et al.* (1995) have shown that the Urick equation (Equation 2.9), written in terms of inverse square sound velocity, takes a particularly convenient form:

$$\frac{1}{v^2} = \frac{1}{v_1^2}\left(1 + \alpha\phi_{Urick} + \delta\phi_{Urick}^2\right), \tag{4.54}$$

where v is the sound velocity measured in the mixture, v_1 is the sound velocity in the pure continuous phase, and ϕ_{Urick} is the volume fraction, determined by applying the Urick equation 2.9 to the sound velocity v determined in the mixture. The parameters α and δ are related to the densities and compressibility of the two phases,

$$\alpha = \left[\frac{\Delta\kappa}{\kappa_1} + \frac{\Delta\rho}{\rho_1}\right]; \quad \delta = \left[\frac{\Delta\kappa}{\kappa_1} \times \frac{\Delta\rho}{\rho_1}\right], \tag{4.55}$$

where $\Delta\kappa = \kappa_2 - \kappa_1$ and $\Delta\rho = \rho_2 - \rho_1$. κ is compressibility, ρ is density, and the subscripts $_1$ and $_2$ refer to the continuous and dispersed phases, respectively. Thus, the quantities which characterize the Urick equation are the fractional differences between the two phases, $\Delta\kappa/\kappa_1$ and $\Delta\rho/\rho_1$ of the compressibility and density, respectively. The Urick equation is therefore a result of scattering theory.

4.3.9.2 The Multiple Scattering Result

If we assume that signal attenuation is small, which is a normal condition for making any measurements in processes, then the sound velocity as determined from multiple scattering theory (Waterman and Truell, 1961; Fikioris and Waterman, 1964; Lloyd and Berry, 1967) can be written in a very similar form to Equation 4.54:

4.3 SCATTERING THEORY

$$\frac{1}{v_2} = \frac{1}{v_1^2}\left(1 + \alpha_e \phi_{scatt} + \delta_e \phi_{scatt}^2 + O(\phi_{scatt}^3)\right), \quad (4.56)$$

where ϕ_{scatt} is the volume fraction corresponding to the sound velocity determined by scattering theory. This presents the possibility of a direct experimental determination of the scattering coefficients from sound velocity measurements as a function of concentration. An experimental plot of $1/v^2$ against ϕ permits the determination of α_e and δ_e. The use of a Urick equation form for analyzing sound velocity data can now be viewed as equivalent to the determination of *effective* Urick equation coefficients α_e and δ_e in which *effective* values for the parameters $\Delta\kappa/\kappa_1$ and $\Delta\rho/\rho_1$ are obtained:

$$\frac{\Delta\kappa_{eff}}{\kappa_1} = \frac{\alpha_e}{2}\left[1 + \sqrt{1 - \frac{4\delta_e}{\alpha_e^2}}\right] \quad (4.57)$$

$$\frac{\Delta\rho_{eff}}{\rho_1} = \frac{\alpha_e}{2}\left[1 - \sqrt{1 - \frac{4\delta_e}{\alpha_e^2}}\right] \quad (4.58)$$

Thus, an effective fractional compressibility and density difference between the two phases of the mixture may be determined. These are not the same as the compressibility and density as determined by static methods, and the values determined this way will in general depend on frequency and particle size. The effective parameters are valid over the full range of volume fraction and therefore characterize the emulsion completely with regard to sound velocity. The effective parameters include the effects of any thermal scattering in the system, and no restriction is placed on the thermal and viscous decay lengths.

In the long wavelength limit and in the absence of thermal scattering, the scattering parameters α_e and δ_e are identical to the Urick parameters α and δ. Thus, the unmodified Urick equation (Equation 4.54) accurately represents the sound velocity as determined by scattering theory *in the long wavelength limit when thermal scattering is absent* and when the viscous skin depth is large in comparison with particle size.

4.3.9.3 The Modified Urick Equation

The discussion of the Urick equation (Equation 4.54) has so far ignored the effects of thermal scattering. Urick and Ament (1949) recognized that Equation 4.54 was a special case of scattering theory but omitted thermal scattering. Figure 4.6 shows that the thermal effects can be far from insignificant and therefore need to be accounted for within a modified Urick equation. As pointed out earlier, the thermal effects are contained within the measured scattering parameters α and δ. However, it is necessary to incorporate these effects explicitly if the compressibility of a dispersed phase is to be determined. In this case, the modified Urick equation may be written in the form given in Equation 2.34, §2.4.5, where the assumption that the density difference is small leads to the neglect of a term

$$+6\left(\frac{\Delta\rho}{3\rho}\right)^2 \qquad (4.59)$$

in δ.

A discussion of the use of this equation for the determination of the adiabatic compressibility of solute molecules and dispersed particles, together with examples, can be found in §2.4.5.

The conditions of validity of the modified Urick equation in terms of the thermal and viscous waves may be characterized quantitatively as the condition that the respective decay lengths be much greater than particle size. The modified Urick equation is only valid in the long wavelength limit.

$$\frac{r}{\delta_s} = \sqrt{\frac{r^2\omega}{2v}} \ll 1, \qquad \frac{r}{\delta_t} = \sqrt{\frac{r^2\omega}{2\sigma}} \ll 1. \qquad (4.60)$$

Before embarking on an ultrasound velocity determination of the volume fraction of a dispersed phase, one should to check that this condition holds (see Figure 4.8).

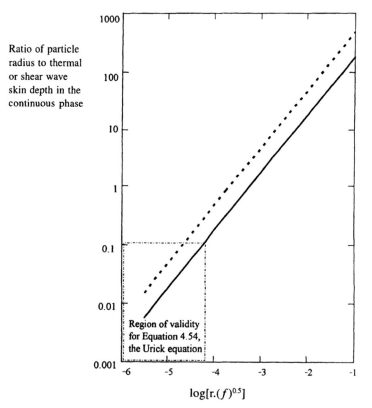

FIGURE 4.8 Ratio of particle size to the thermal (- - -) and shear (—) decay lengths in water at 20°C. Data abstracted from MathCad sheet in the Appendix.

4.3 SCATTERING THEORY

In Figure 4.8 the ratio of particle radius to decay length (also known as *boundary layer thickness* or *skin depth*) has been plotted for the thermal and shear (also known as viscous) waves in water, against the parameter $rf^{0.5}$, which is chosen because the scattering processes under consideration scale according to this parameter, under the limiting conditions considered in this section. From Figure 4.8, it can be seen that at 1 MHz, the modified Urick equation can be applied to particles less than about 0.1 μm. Of course, another version of the Urick equation can be derived for the case when the thermal and shear decay lengths are less than the particle size (i.e., 0.00148 m $\gg r \gg$ 0.1 μm), from the formulas in Tables 4.4 and 4.5.

4.3.9.4 Experimental Determination of the Scattering Coefficients

There are very many merits to determining the scattering coefficients experimentally, rather than calculating them from the explicit formulas or from numerical solutions of the full scattering coefficients. First of all, it is not necessary to know all the physical constants involved in an explicit calculation. Second, very many effects such as particle size distribution, particle shape, variations in the decay lengths of the thermal and shear waves, and overlap effects are all subsumed into the two numbers α and δ.

The coefficients α and δ are related to the modified scattering coefficients, assuming that attenuation and $\Im m(a_n)$ can be neglected, through

$$\alpha = \Re e(a_0 + 3a_1), \quad \delta = \left[3\Re e(a_0)\Re e(a_1) + 6(\Re e(a_1))^2 \right], \quad (4.61)$$

and the scattering coefficients (Equation 4.46) are related to the experimental velocity coefficients when there is a single dispersed component by

$$\Re e(a_0) = -\frac{\alpha}{2}(1 \pm 3\Phi), \quad (4.62)$$

$$\Re e(a_1) = \frac{\alpha}{2}(1 \pm \Phi). \quad (4.63)$$

where

$$\Phi = \sqrt{1 - \frac{4\delta}{3\alpha^2}}. \quad (4.64)$$

Correct correspondence with scattering theory requires the choice of the "minus" solution of these equations. The results just outlined provide a very powerful approach to determining the scattering coefficients. Measurements can be made of the relationship between ultrasound velocity and attenuation, and the dispersed phase concentration or some other independent variable of the system under investigation.

Two approaches may be adopted to the measurement of the scattering coefficients. First, the scattering coefficients may be obtained from Equation 4.56, the complete multiple scattering formulation using ultrasound velocity data. These are the scattering coefficients, not necessarily of individual particles or species, but of the collective properties of the system. These quantities can be simply determined from measured values of α and δ, such as can be found in the previous chapter. The compressibility and density can then be obtained directly from Equations 4.57 and 4.58. This method requires the fewest assumptions. However, the compressibility determined this way will include the effects of thermal scattering.

Second, ultrasound velocity data can be used to determine the "scattering coefficients" through Equations 4.62 to 4.64 above. Experimental evidence for the validity of scattering theory extends to particle sizes as small as 130 nm (Schröder and Raphael, 1992), and it seems reasonable to apply it to solute molecules in solution. The heat flow between the particle and the solvent is related to the ratio R (Equation 2.34), the difference in temperature between the substances that make up the suspension. R can be negative, corresponding to heat flow between solute and solvent in antiphase to the pressure fluctuation. If the scattering particles are molecules, then each molecule will be subject to the combined thermal fields of its neighbors, even at extremely low concentrations by volume. The solute parameters will also be altered in nonideal solutions, by hydration (Harned and Owen, 1958) and by the heat of mixing (Larkin, 1975). In the case of a mole fraction of ethanol mixed with water of 0.05, the scattering coefficients determined by the method outlined in §2.4.5 are given in Table 4.6.

The thermal term, θ, determined from the linear coefficient α, is very different from θ', which is determined from the quadratic coefficient δ. This is consistent with concentration dependent multiple scattering of the thermal wave, which introduces an additional component into the quadratic term, corresponding to a change in the ratio of the specific heats, which may be concentration dependent. The calculation of the scattering coefficients is therefore based on the linear term only.

The approach to the determination of the scattering coefficients based on the modified Urick equation, Equation 2.34, permits the effects of thermal scattering to be accounted for explicitly. The thermal scattering contribution can be separated out of the adiabatic compressibility determination in Equation 2.34, as shown earlier, permitting the properties of solute molecules to be compared, without the added complication of accounting for thermal scattering in the solution in which they happen to be measured. If this is not a consideration, then Equation 4.56 should be used.

TABLE 4.6 The Real Parts of the Zero- and First-Order Modified Scattering Coefficients[a]

T	α	δ	c	θ_{exp}	θ'	R_{exp}	γ	S	$Re(a_0)$	$Re(a_1)$
0	−1.68	2.03	0.0032	0.18	−12.7	25.04	1.00057	0.195	0.348	−0.675

[a] Determined from velocity of sound and density data of D'Arrigo and Paparelli (1987), other data is taken from Table 2.3.

4.3 SCATTERING THEORY

In conclusion, the scattering formulation is applicable to solutions. The compressibility determined from sound velocity measurement under the usual conditions is adiabatic, but the thermal term must be included and can be very significant (Table 2.3). The data in Table 2.3 indicate that the very large number-density of scatterers in solutions, even at low mole fraction or volume fraction, causes additional scattering effects which are not yet included in scattering theory.

4.3.10 MULTIPLE DISPERSED PHASES

Very many systems are composed of more than one dispersed phase. Examples include cosmetic preparations comprising water (continuous phase), oil (dispersed), wax (dispersed), and polymer (dispersed) and food emulsions such as mayonnaise: water (continuous), oil (dispersed), polymer (e.g., lecithin, dispersed). Both these examples are in fact even more complicated; for example, the different phases in the cosmetic will have very different particle size distributions. Nevertheless, scattering theory can be applied to such complex systems (McClements et al., 1990c), with the help of appropriate approximations and choice of limiting cases. Scattering theory permits the development of a general theory for mixed systems, which appears in Pinfield (1996) and is reproduced here.

The following equations are based on the scattering theory described earlier. Each component in the mixture must separately satisfy the relevant approximations and limiting conditions as they appear in the earlier discussion. The resulting theory does not include interactions between the various components of the mixture, such as might occur in flocculation or other aggregation processes. The effects of flocculation and aggregation will be discussed later.

In the following treatment, the imaginary parts of the modified scattering coefficients (Equation 4.46) have been neglected. Each component must act independently of the other components during the passage of the sound wave and must be randomly distributed. For each component, the velocity is given by Equation 4.56 and the result for the combined system is obtained by summing the coefficients for each phase in proportion to their number density. The combined coefficient is obtained in the same way as for a polydisperse sample (Equation 4.45), each size fraction is treated as a separate scattering component. The result for the velocity of sound is

$$\frac{1}{v^2} = \frac{1}{v_1^2}\left[1 + \sum_i (\phi_i \operatorname{Re} a_{0i} + 3\phi_i \operatorname{Re} a_{1i})\right.$$
$$\left. + 3\left\{\left(\sum_i \phi_i \operatorname{Re} a_{0i}\right)\left(\sum_i \phi_i \operatorname{Re} a_{1i}\right) + 2\left(\sum_i \phi_i \operatorname{Re} a_{1i}\right)^2\right\}\right] \quad (4.65)$$

where i represents each dispersed phase species. This equation can be simplified through the use of the relationship between the individual scattering coefficients and the velocity equation for each phase separately (Equations 4.56 and 4.61) to give

$$\frac{1}{v^2} = \frac{1}{v_1^2}\left(1+\sum_i(\alpha_i\phi_i+\delta_i\phi_i^2)+\sum_i\sum_{j>i}\Gamma_{i,j}\phi_i\phi_j\right), \quad (4.66)$$

where

$$\Gamma_{i,j} = 3(\operatorname{Re}a_{0i}\operatorname{Re}a_{1j}+\operatorname{Re}a_{0j}\operatorname{Re}a_{1i}+4\operatorname{Re}a_{1i}\operatorname{Re}a_{1j}). \quad (4.67)$$

To obtain these equations, multiple scattering events have been limited to double scattering effects only, i.e., second-order in concentration. The final term in Equation 4.66 is the only one which cannot be determined directly from experiment in systems comprising a single dispersed phase. If the scattering coefficients can be written in the form of Equations 4.62 and 4.63, then the extra term can be simplified

$$\Gamma_{i,j} = \frac{3}{2}\alpha_i\alpha_j(1-\Phi_i\Phi_j), \quad (4.68)$$

where Φ is given by Equation 4.64.

If the density difference between the dispersed and continuous phase is small and the term

$$+6\left(\frac{\Delta\rho}{3\rho}\right)^2$$

(Equation 4.59) can be neglected, then the above Equation 4.68 becomes

$$\Gamma_{i,j} = \frac{1}{2}\alpha_i\alpha_j(1-\Phi_{Ur,i}\Phi_{Ur,j}) \quad (4.69)$$

$$\Phi_{Ur,i} = \sqrt{1-\frac{4\delta_i}{\alpha_i^2}}. \quad (4.70)$$

The following equation for the attenuation cannot be simplified so easily:

$$\alpha_s = \sum_i(\chi_i\phi_i+\varsigma_i\phi_i^2)+\sum_i\sum_{j>i}\zeta_{i,j}\phi_i\phi_j, \quad (4.71)$$

where χ is the first coefficient of the attenuation-concentration relationship, ζ is the cross scattering coefficient in the attenuation, and ς is the second coefficient of the attenuation–concentration relationship.

4.3.11 MATHCAD CALCULATION RESULTS

The description in the previous section applies over the whole range of volume fraction, unlike previous descriptions which employed the effective density and compressibility, as opposed to their *fractional* differences.

The explicit solutions of scattering theory which appear in earlier sections (§4.3.8 to §4.3.10) are valid in the long wavelength limit. These solutions are suitable for use in MathCad spreadsheets, and an example of such a sheet is given in the

4.3 SCATTERING THEORY

Appendix. The calculations of scattering and the figures for the frequency and size dependence of the scattering used are based on the equations in this MathCad sheet. Where appropriate, a data table will be given, but the calculation scheme remains identical to that in the Appendix.

In Figures 4.9 to 4.11 the sound velocity, compressibility, and attenuation are plotted against the parameter $f^{0.5}r$, for a sunflower oil-in-water emulsion. The parameter $f^{0.5}r$ is used because the propagation parameters above scale with this parameter, in the long wavelength limit. This is the same system as has been plotted in Figures 4.6 and 4.7, but the data are for 20°C and MathCad has been used together with explicit solutions for the scattering coefficients, rather than the numerical methods used earlier in this chapter. The contributions from the thermoelastic (compressibility) and viscoinertial (density) terms are plotted separately, together with their combined contribution. In this case the scattering is dominated by the thermoelastic term. The data are taken from Table 4.7.

These curves display the features which have already been described in Figures 4.6 and 4.7. The plot for the thermal contribution has been obtained by assuming that the density term is negligible.

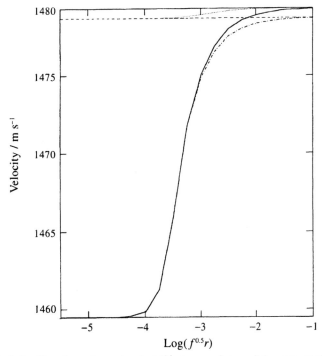

FIGURE 4.9 Plot of sound speed against $f^{0.5}r$ for a sunflower oil-in-water emulsion at 20°C. The solid line (———) is the prediction for the overall sound speed in the presence of oil droplets. The contributions from the thermal (– · – · – ·) and viscous (·······) terms and the unmodified Urick equation (- - - -) are also shown. The data on which this calculation is based are given in the text.

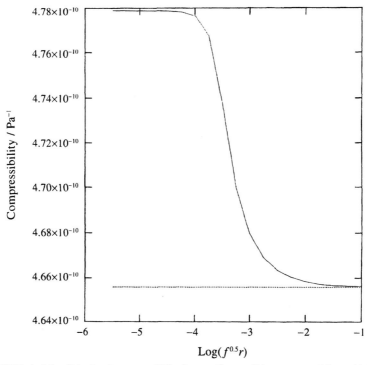

FIGURE 4.10 Calculated compressibility in a sunflower oil-in-water emulsion at 20°C. ········· compressibility in the absence of thermal scattering. The MathCad sheet in the Appendix was used here, and the data are from Table 4.7.

TABLE 4.7 Thermophysical Properties of Component Phases[a]

Property at 20°C	Water	Sunflower oil
Velocity (m s^{-1})	$v_1 := 1482$	$v_2 := 1470$
Density (kg m^{-3})	$\rho_1 := 998$	$\rho_2 := 920$
Attenuation (Nepers m^{-1})	$\alpha_1 := 0.4$	$\alpha_2 := 1.7$
Viscosity (Pa s)	$\eta_1 := 0.001$	$\eta_2 := 0.054$
Specific heat (J kg^{-1} m^{-3})	$C_1 := 4182$	$C_2 := 1980$
Thermal conductivity (W m^{-1} K^{-1})	$\tau_1 := 0.591$	$\tau_2 := 0.17$
Coefficient of cubical expansion	$\beta_1 := 0.00021$	$\beta_2 := 0.00071$
Compressibility	$\kappa_1 := \dfrac{1}{v_1 \cdot v_1 \cdot \rho_1}$	$\kappa_2 := \dfrac{1}{v_2 \cdot v_2 \cdot \rho_2}$
Volume fraction	$\phi := 0.2$	$(1-\phi) = 0.8$
Radius		$r := 1 \cdot 10^{-6}$

[a] Data for Figure 4.9.

4.3 SCATTERING THEORY

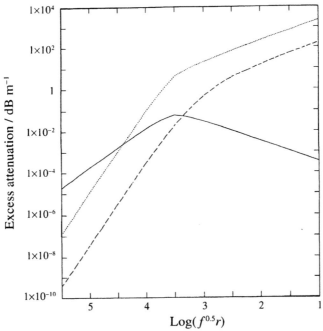

FIGURE 4.11 Calculated excess attenuation per meter for a sunflower oil-in-water emulsion at 20°C. In addition to the total excess attenuation (—), the contributions from the thermal (·····) and viscous terms (– – –) are shown.

Note that in the case of sunflower oil-in-water, the contribution of the viscoinertial effect is very small in comparison to the thermal effect (the attenuation in Figure 4.11 is plotted on a log scale). Note also that the unmodified Urick equation gives an incorrect sound velocity in the long wavelength limit, as seen earlier. This is because thermal effects modify the apparent compressibility in the long wavelength limit, particularly when $k_t r \ll 1$. As the frequency gets smaller and/or the particle size reduces, the sound velocity becomes frequency independent and the modified Urick equation (Equations 4.56 to 4.58) applies.

Figure 4.10 shows the effective compressibility plotted against the parameter $f^{0.5}r$. It can be seen that in this case the compressibility in the long wavelength limit is higher than the volume-averaged value, in accordance with Equation 4.26. In the case of compressibility, it is at higher frequencies and particle sizes that the volume-averaged values are correct.

4.3.12 EXPERIMENTAL VALIDATION OF ACOUSTIC SCATTERING THEORY

There now exists a sizeable body of evidence in support of the acoustic multiple scattering theory just outlined. Apart from Epstein and Carhart (1953), who studied fogs, the first thorough experimental validation of single-particle scattering theory was carried out by Allegra and Hawley (1972) on hexadecane oil-in-water emulsions and on aqueous suspensions of polystyrene spheres. Some workers (Holmes and Challis, 1993; Holmes *et al.*, 1993, 1994) have found significant discrepancies between the predictions of this theory and experimental results on suspensions of polystyrene spheres, especially in frequency and particle size dependence. However, this is due either to the violation of some of the fundamental assumptions of the theory, such as the absence of particle interactions, or to poor characterization of the physical properties of the dispersed phase; or a combination of both. Barrett-Gültepe *et al.* (1983) observed similar deviations for polystyrene latices which they attributed to a decrease in the compressibility of the polystyrene latex due either to hydrophobic interactions or modification of scattering by a PVA coating on the particles. Holmes *et al.* (1993), who appear to have been unaware of the earlier work by Barrett-Gültepe *et al.* (1983), suggest that the differences may be due to experimental uncertainties in the many physical constants required for a scattering calculation. Holmes *et al.* (1993) analyzed the effect of the uncertainties in some of the parameters on the calculation of ultrasound attenuation and velocity and found significant effects in some cases.

It is certainly much more difficult to make accurate attenuation measurements than it is to measure velocity accurately, and this fact has greatly influenced the presentation in this book. Experimental validation of acoustic multiple scattering theory is summarized in Table 4.8, which is not a comprehensive account of all work carried out using scattering theory to interpret experimental data. In particular, a large body of work comprising studies of solid dispersions in liquids, solid-in-solid dispersions, and liquid/gas dispersions in solids has been omitted. For example, geophysicists such as Kuster and Toksöz (1974a, b) have used forms of scattering theory to interpret seismic data. The interested reader is referred to the reviews by Harker and Temple (1988, 1992), Kytömaa (1995), McClements (1991), and Riebel and Löffler (1989) for more details. Harker and Temple (1988) present an alternative method to that of scattering theory which they call a "coupled-phase' model". There is evidence that the coupled-phase model gives more accurate results for the attenuation at high concentrations than scattering theory, and it has been noted elsewhere that scattering theory overpredicts the attenuation at high volume fractions (McClements, 1992; Evans and Attenborough, 1996). A form of coupled-phase model, developed by Biot (1956a, b), is appropriate for systems where one continuous or semicontinuous phase interpenetrates another, permitting relative motion between the two. An example of such a system would be a sponge containing water. Coupled-phase theories are discussed further in §5.2.1.

The summary in Table 4.8, indicates that very many experimentalists have both felt enough need to employ scattering theory, and had enough confidence in the quality of the work of their predecessors, to make the effort to use it. It is also clear that a wide range of apparatuses have been assembled in laboratories over the years, but that it still requires a considerable effort to assemble the equipment required to carry out measurements over a range of frequencies. This is still the case, although the situation may alter in the near future.

4.3 SCATTERING THEORY

A glance at the table also shows that both pulse and continuous-wave techniques have been used. This is because each has its own merits. The pulse technique is easily automated and rapid, costs less to assemble, and is less demanding of the engineering of the ultrasound cell. On the other hand, continuous-wave techniques can be much more accurate. There are good reasons to think that apparatuses can be constructed which share the benefits of both techniques. This subject will be returned to in the sections on particle sizing in the next chapter.

TABLE 4.8 Experimental Vindication of Acoustic Scattering Theory[a]

Experiment	Bandwidth	Material	Authors
Attenuation, reverberation chamber	500 to 8000 Hz	Water fog	Epstein and Carhart (1953)
Attenuation and velocity, pulse and standing-wave methods	4 kHz to 600 kHz	Kaolinite, sand, kaolinite/sand mixtures and soil	Hampton (1967)
Attenuation, pulse method	9 MHz to 165 MHz	Hexadecane oil-in-water, toluene in water, polystyrene latex 44 nm to 653 nm	Allegra and Hawley (1972)
Attenuation, pulse method	4 to 50 MHz	Polystyrene spheres and aggregation of erythrocytes in human blood	Horak (1972)
Attenuation and velocity, pulse method	250 kHz	Polystyrene-in-oil, polystyrene-in-water (did not take thermal scattering into account)	Kuster and Toksöz (1974b)
Attenuation, standing wave technique	0.6 to 10 MHz	Dilute suspensions of rayon, polyester, and nylon fibers (7 to 120 µm)	Habeger (1982)
Velocity by interferometer	2 MHz	Polystyrene-in-water	Barrett-Gültepe et al. (1983)
Velocity and attenuation, pulse method	5 MHz	Glass spheres (37 to 90 µm) in lubricating and hydraulic fluids	Harries (1985)
Attenuation and velocity, pulse method	100 kHz to 185 MHz	Perfluorochemical emulsions, 10 to 1000 nm	Barrett-Gültepe et al. (1988)

(continues)

TABLE 4.8 (continued)

Experiment	Bandwidth	Material	Authors
Attenuation and velocity, pulse method	100 kHz to 185 MHz	Coal–water dispersion stability	Barret-Gültepe et al. (1989)
Attenuation, standing-wave technique	1.7 to 81 MHz	Glass beads, quartz sand and polyamide fiber, 19 to 906 μm.	Riebel and Löffler (1989)
Attenuation and velocity, pulse method	1.25 to 10 MHz	Oil droplets (0.55 to 10.2 μm) in salad cream	McClements et al. (1990a)
Attenuation and velocity, standing wave method	5 to 96 MHz	Octadecane oil-in-water emulsions, both crystallized and uncrystallized, 0.2 to 30 μm	Javanaud et al. (1991) Robins et al. (1991)
Attenuation and velocity, pulse method	0.5 to 4 MHz	Silicon carbide (1.5 to 3.25 μm) in water and in ethylene glycol	Harker et al. (1991)
Attenuation and velocity, pulse method	0.2 to 7 MHz	Hexadecane oil-in-water emulsions (0.1 to 1.8 μm, $\phi < 0.13$)	McClements (1992)
Attenuation, standing wave technique	0.5 to 100 MHz	Titanium dioxide (0.3 to 0.6 μm) and glass spheres. Claims 0.1 to 1000 μm, but no data supporting the full range	Alba (1992)
Attenuation, standing wave (400 kHz < f < 1 MHz) and pulse method	400 kHz to 100 MHz	Silicon oil-in-water emulsions, 132 to 206 nm	Schröder and Raphael (1992)
Attenuation and velocity by pulse method	2 to 60 MHz	Polystyrene suspensions (200 to 611 nm), poly(methylmethacrylate) suspension, cyanazine slurry	Holmes and Challis (1993)
Attenuation and velocity by pulse method	2 to 60 MHz		Holmes et al. (1993)
Attenuation and velocity by pulse method	2 to 60 MHz		Holmes et al. (1994)
Attenuation by pulse method	600 kHz to 2 MHz	Water-in-alkyd-resin emulsions, 0.5 to 3.5 μm	Aurenty et al. (1995)

(continues)

TABLE 4.8 *(continued)*

Experiment	Bandwidth	Material	Authors
Velocity and concentration by pulse method	2.25 MHz	Oil-in-water emulsions containing polymer	Pinfield et al. (1994)
Attenuation by standing wave pulse method	2 to 50 MHz	Submicron slurries below 5 vol% dispersed phase	Scott and Boxman (1995)
Velocity and concentration by pulse method	2.25 MHz	Oil-in-water emulsions containing polymer	Pinfield et al. (1995)
Velocity and attenuation by pulse method (McClements and Povey, 1989)	1.25 MHz	Sunflower oil-in-water emulsion	Evans and Attenborough (1996)
Velocity and concentration by pulse method	2.25 MHz	Oil-in-water emulsions containing polymer	Pinfield (1996)

[a]The pulse method is discussed in detail in Chapter 2 and Chapter 5; the continuous wave method is discussed in Chapter 5.

4.4 SCATTERING FROM BUBBLES

Bubbles are ubiquitous in industrial processes and in the laboratory. Since bubbles have a profound impact on acoustic propagation, it is necessary to remove them before other acoustical effects can be studied properly (Chapter 2). However, the very sensitivity to bubbles and richness of phenomena in their presence makes acoustics potentially a powerful technique for the study of bubbles.

A formal solution of the problem of scattering from bubbles differs from the weak scattering case thus far considered for at least three reasons. First the bubble in a liquid medium such as water is a strong scatterer of sound, primarily because of the large density difference between the bubble and its surroundings. This greatly complicates the multiple scattering problem because it is no longer possible to assume that the amplitude of the scattered wave can be neglected. Second, the bubble exhibits resonant behavior in the presence of an acoustic field which greatly complicates the single-particle scattering problem. Third, the surface energy of the bubble is a very significant part of the total energy of small bubbles and more than one mode of oscillation of the bubble may be important, so that the single-particle scattering coefficients can no longer be determined, in the general case, by a simple superposition of the single-frequency solutions. The result is that general solution of the scattering problem for a cloud of bubbles is far more troublesome than for a dispersion of solid or liquid particles in a liquid medium. It is not possible to do justice to this complex and much-studied area in this short section. Instead, the basic physics will be briefly outlined, the main results stated, and the reader guided to further reading.

The study of bubbles has been greatly stimulated by the development of sonar for underwater sensing and communications. Since the First World War, the detection of bubble clouds produced by submarine propellers has been used as a means of submarine detection (Domenico, 1982). Acoustic propagation in the sea is greatly affected by bubbles, and bubbles are extremely effective acoustic screens. Indeed, bubble clouds have been used to protect underwater structures from the effects of blast. A blast wave in water may be viewed as an acoustic wave, which may be reflected by a cloud of bubbles of the correct size.

The first significant study of bubble acoustics is due to Minnaert (1933), who wrote a paper entitled "On Musical Air Bubbles and the Sounds of Running Water." This beautiful and easily understandable work explained the sounds of running water in terms of bubbles trapped within the water, oscillating at a resonant frequency now called the Minnaert frequency (ω_M). Anyone interested in this subject should start with this paper, which captures the essence of the physics of bubble resonance.

$$\omega_M = \sqrt{\frac{3\gamma p}{\rho r^2}} \qquad (4.72)$$

Bubble resonance arises primarily because of the balance between the pressure inside the bubble and the restoring force which arises from the displaced volume of liquid surrounding an expanding bubble; the resulting pulsation which arises from any perturbation of the bubble will have the frequency of Equation 4.72. This equation is a useful starting point for "back-of-the envelope" calculations of the likely effects of bubbles on acoustic propagation. However, it is likely to be in error for small bubbles (large bubbles rapidly cream out of a liquid). The major deficiencies of Equation 4.72 are, first, that it does not account for the surface energy of the bubble, which gives rise to surface tensional forces. The surface energy becomes an increasingly important part of the total energy of a bubble as it gets smaller and will be very sensitive to the state of the interface between the bubble and the surrounding liquid (Commander and Prosperetti, 1989). Second, it does not account for the different modes of oscillation of which a bubble is capable. In addition to undergoing the usual pulsation, which is spherically symmetric, more complicated oscillations may occur, rather like the different modes of which a drum surface is capable of (Chivers et al., 1992). Finally, the effects of multiple scattering profoundly alter the single-scattering behavior (d'Agostino and Brennen, 1988), and the single bubble resonance frequency can only indicate the center of a resonance that is usually several decades of frequency wide (see Figure 4.12).

The subject has developed very much in terms of refining the physics of the oscillation of a single bubble, and landmarks in its theoretical development include work by Nishi (1975), Commander and Prosperetti (1989), Varadan et al. (1985), and Gaunaurd and Überall (1981). The interested reader is referred to the

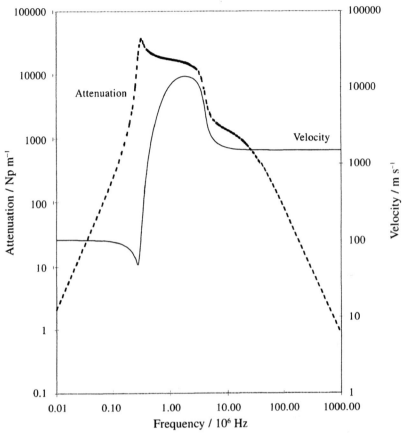

FIGURE 4.12 Velocity and attenuation as a function of frequency for bubbles of 10-μm diameter and a concentration of 2% v/v (Fairley, 1992; Fairley et al., 1991).

comprehensive review by Leighton (1996) on the subject of the "acoustic bubble" and the review by Commander and Prosperetti (1989) for a detailed account of the physics involved.

In Figure 4.12 is presented the results of a calculation of bubble resonance based on the theory of Gaunaurd and Überall (1981). Although this theory contains many approximations and later, more sophisticated treatments are more accurate, it has the great merit of relatively simple computer implementation. In addition, the theory reproduces most the main features of acoustic bubble resonance. A further merit to this treatment from the point of view of this work is that much of the mathematical apparatus of scattering theory in this chapter is used to obtain the result for bubble resonance. At low volume fractions this theory yields results which agree well with experiment, although it does not fare well at intermediate and high volume fractions (Fairley et al., 1991). Strictly, it is not the volume fraction that matters, but the bubble number density, since it is the number of bubbles in the acoustic field which determine whether multiple scattering is likely to be dominant. For this theory to fail, it is only necessary for there to be more than a handful of bubbles in the acoustic beam. Therefore, for small bubbles, a "low" volume fraction may be extremely low indeed, whereas for large bubbles, a low number density may mean a very high volume fraction.

The Gaunaurd and Überall theory proceeds according to the plan of Nishi (1975), who calculates the modified single-particle scattering coefficients (a_n) along the lines outlined in the previous sections of this chapter, suitably modified for a bubble suspended in a liquid. The single-particle scattering coefficients are then combined via a radial distribution function (Twersky, 1978; Tsang *et al.*, 1982; Ma *et al.*, 1990) to yield a dispersion relation. This relation must be solved numerically for bubbles, and the result depends critically on the form of the radial distribution function. At high number densities, bubbles act as an ensemble (d'Agostino and Brennen, 1988), and the bubble cloud as a whole can have a series of resonances which give rise to additional scattering.

There are three important features to note from Figure 4.12. First, the resonance is wide, extending over nearly two frequency decades in the case of velocity. Second, in the case of the velocity of sound, there are two regions in which the velocity ceases to display frequency dependence, at high frequencies and at low frequencies. The low-frequency behavior is given by the modified Urick equation, with the appropriate single-particle scattering coefficients, and the low velocity is to be expected (see Chapter 2). At high frequencies, the velocity is independent of bubble concentration. (Bear in mind that these results are only valid for low number densities of bubbles.) Third, the attenuation at resonance is high enough at most bubble concentrations to make measurements in standard transmission apparatus impossible. Therefore, it is usually necessary to study the acoustics of bubbly systems using backscattering or reflection coefficient methods. The application of these results to systems containing bubbles will be considered in Chapter 5.

5

ADVANCED TECHNIQUES

5.1 PARTICLE SIZING

5.1.1 INTRODUCTION

In the previous chapter, analytical techniques for quantifying acoustic scattering were reviewed, providing a firm theoretical basis for ultrasonic particle sizing. The effects of weak scattering on the acoustic wave were analyzed in detail; the derived scattering coefficients may be determined from experimental measurement of the amplitude and phase of the forward-scattered acoustic signal. In principle, measurements such as these can be used to determine, among other things, particle size distribution, dispersed phase volume, and the adiabatic compressibility of a dispersed phase. In addition, these measurements may be used to determine properties of the continuous phase through measurement of relaxation and other properties of the dispersed phase such as its critical micelle concentration and solubility (§5.1.4).

There is a need for a particle sizing technique which is nondestructive and noninvasive, and that can size colloidal dispersions of a few tens of nanometers upwards. Light-scattering techniques can be noninvasive but multiple scattering becomes a problem in many systems at concentrations above a few fractions of a percent (Yan and Clarke, 1989). As a result, routine particle size measurement by laser light-scattering normally involves sample dilution. In addition, samples are often sonicated to overcome any aggregation effects. NMR may be used for indirect particle sizing, by following the creaming or sedimentation of the dispersed phase through measurement of the T1 decay (Pilhofer and McCarthy, 1993), and ultrasound may be used in a similar way (Carter *et al.*, 1986; §3.5). Ultrasound analysis has been identified (Roberts, 1996) as having a number of potential advantages over conventional light-based techniques:

1. It is easier to account for multiple scattering effects, and consequently the technique can be used over a very wide range of concentrations, up to the random monodisperse close packing limit of 60% and beyond. The lower limit depends on a number of factors, but could be below 0.1%. Ultrasound measurements can generally be made in materials which are opaque to light, provided air bubbles are absent.

2. Ultrasound instrumentation can be made robust enough for use in on-line measurement in a factory, as well as for the laboratory bench.

3. A very wide range of particle sizes can be determined using the technique. A dynamic range of 0.01 to 1000 μm is claimed for the Ultraspec instrument (Roberts, 1996).

4. Measurement of both phase (velocity) and amplitude (attenuation) is available, increasing the accuracy of particle size distribution determinations.

5. Larger particles (taking measurements beyond the long wavelength region) can be sized by modifying scattering theory along the same lines as are followed for Light-scattering. First, a Rayleigh–Gans–Debye (Yan and Clarke, 1989) approximation can be used in which a form factor accounts for scattering from different parts of the molecule or particle, as the geometric limit is approached. Second, Mie theory (Yan and Clarke, 1989) may be invoked to account for scattering in the geometric limit. This is beyond the scope of this book.

The range of ultrasound particle measurement depends on a host of factors involving both the instrument (frequency range, sensitivity) and the sample (sample properties, particle size distribution). An examination of Figures 4.6 and 4.7 also shows that there are circumstances where particle sizing is not possible because the velocity and attenuation are independent of particle size. These problems are reduced by increasing the dynamic range of the instrument itself; for example, the Ultraspec instrument operates between 1 and 200 MHz. An ultrasound dynamic range this wide requires the use of at least two transducer pairs, one for low frequencies (normally a ceramic piezoelectric) and one for high frequencies (quartz or lithium niobate crystals operating in thickness resonance mode). Poly(vinylidene difluoride) (PVDF) or block copolymer transducers are likely to be used increasingly in future, as they become more widely available, because of their wide and flat frequency response. Frequencies below 1 MHz, which may be desirable for accurate sizing of larger particles, normally require larger sample cells because of the increased aperture of the transducer and the longer wavelength of sound.

5.1.2 REVIEW

The subject of ultrasound particle sizing has been reviewed by Kytömaa (1995). This is a very rapidly developing area, and there are a number of alternatives to the scattering theory approach adopted here. In addition, there have, over the years, been a number of empirical and semiempirical attempts at using ultrasound for particle sizing, particularly using ultrasound tomography (Kitamura *et al.*, 1995; Mahony, 1987; Mahony *et al.*, 1989; Rokhlenko, 1986).

There are several approaches to the determination of particle size: pulse and continuous-wave measurement of amplitude and phase, electro-sonic amplitude (ESA) (O'Brien, 1988, 1990; Wade *et al.*, 1995; O'Brien *et al.*, 1994, 1995), ultrasound vibration potential (UVP) (Dickinson and McClements, 1995b), and measurement of creaming rate (Carter *et al.*, 1986). Sometimes different methods can be used simultaneously within a single apparatus.

Broad-band methods of determining particle size have been published by Javanaud *et al.* (1986); Barrett-Gültepe *et al.* (1988) (perfluorochemical emulsions); Barrett-Gültepe *et al.* (1989) (coal–water dispersions); Riebel and Löffler (1989) (glass beads and sand); McClements *et al.* (1990c) (a food emulsion —salad cream); Alba (1992) (titanium dioxide suspensions); Schröder and Raphael (1992) (130-nm silicon oil-in-water emulsions); de Boer *et al.* (1995); and Bouts *et al.* (1995) (an empirical technique for following the aggregation of asphaltenes). This is clearly an active area in which a great deal of further development is likely. In 1988, Barret-Gültepe *et al.*, sized particles between 20 and 1000 nm, using attenuation and velocity measurements between 100 kHz and 185 MHz. The Ultraspec ultrasound particle sizer has been developed from Alba's patent (1992), which is based on the analysis of attenuation data using scattering theory as described in Allegra and Hawley (1972). Roberts (1996) states that the Ultraspec is based on a continuous-wave technique. Many of the measurements just referred to have used ultrasound attenuation measured as a function of frequency to determine particle size (for more details, see Table 4.8) and have neglected to measure the velocity against frequency.

Rather than examine each author's work in detail, general points relevant to the future development of ultrasound particle sizing will be addressed.

5.1.3 THEORETICAL LIMITATIONS OF ACOUSTIC PARTICLE SIZING

Acoustical particle sizing depends on a contrast in properties between the dispersed and continuous phases of the suspension under investigation, just as Light-scattering does. If this contrast is absent, then the particles will be undetectable acoustically. Where Light-scattering methods are concerned, users are familiar with the need to measure or look up the refractive index of the material they wish to size. Similar procedures are essential in acoustic particle sizing. In water-containing systems, it is essential to take the temperature dependence of the velocity of sound into account. The differing signs of the temperature coefficient of velocity between water and most other materials can give rise to complicated temperature-dependent effects (see §2.2.4, §3.2.4, §3.4.3; Figure 3.10). The existence of two regions where the velocity of sound is independent of particle size and one region where the attenuation is independent of particle size is a further complication. However, combined measurements of velocity and attenuation can help to overcome this problem. It is worth noting that successful acoustical sizing depends on a low attenuation of the acoustic signal in the sample, just as Light-scattering depends on the ability to transmit light through a sample. Consequently, a highly attenuating liquid medium will be a barrier to successful acoustic sizing, although backscattering and reflectance coefficient measurements may give some guide to particle size at the surface. In some circumstances, where concentrated dispersions are concerned, scattering can give rise to high total attenuation. Once this occurs, acoustical sizing cannot be carried out with present methods. It is worth pointing out, however, that sound energy is still present in the dispersion in incoherent form (Figure 2.14). The development of an incoherent detector of ultrasound would greatly extend the range of applicability of acoustic particle sizing. In particular, sizing of bubbly and voided systems would become viable.

5.1.4 RELAXATION EFFECTS

Ultrasonic particle sizing depends on a fit of a model of the dependence of signal amplitude and phase as a function of frequency to the measured data. This model presumes that the frequency dependence arises solely from the interface between the dispersed phase and the continuous phase. It is quite possible for frequency-dependent effects to be intrinsic to the continuous and to the dispersed phase. In this case, it is necessary to remove the material-intrinsic frequency-dependent contribution before analyzing the ultrasound data to obtain particle size and particle size distribution. In the case of attenuation, this is done by measuring the frequency-dependent attenuation in the pure phase and subtracting the result from the total attenuation (Equation 4.40). The frequency dependence of the pure phases should always be measured if possible.

Frequency-dependent attenuation may also arise from interaction between the continuous and dispersed phases, as in the case of exchange of molecules between the two which occurs, for example, in micelles between the micelle and the solvent (Borthakur and Zana, 1987; Frindi *et al.*, 1991,1994a,b).

The pure-phase frequency-dependent data may contain very valuable information about the chemistry and physical chemistry of the material. There is already a large body of work (Borthakur and Zana, 1987) employing ultrasound relaxation techniques between 0.5 MHz and 80 MHz to study the exchange of surfactant monomer between micelles and solvent. This may be used to determine the critical micelle concentration, a key parameter in the characterization of surfactants. A discussion of this work is beyond the scope of this book, but it lies in the experimental tradition of chemical relaxation and absorption studies pioneered by workers such as Wyn-Jones (Tiddy *et al.*, 1982), Heasell (Heasell and Lamb, 1956), Lamb (1969), Eggers and Funck (1973), and Slutsky (1981). An alternative approach to the relaxation time analysis of ultrasound data has been adopted in this work, based on scattering theory. However, Edmonds (1981) and Pethrick (1983) give good summaries of the relaxation time approach to analyzing data from polymer solutions.

5.1 PARTICLE SIZING

If only one relaxation process is present, then a very distinctive dependence of attenuation on frequency is expected (Mason et al., 1964–1992), of the form

$$\alpha_r = \frac{C_r}{2v\omega_r} \frac{\frac{\omega^2}{\omega_r^2}}{1+\frac{\omega^2}{\omega_r^2}}, \quad \omega_r = \frac{2\pi}{\tau_r}, \quad (5.1)$$

where α_r is the attenuation due to relaxation; ω is radial frequency; ω_r is relaxation frequency; C_r is coupling constant; v is velocity of sound; and τ_r is relaxation time.

The combination of velocity and attenuation measurements made as a function of frequency (§5.1.5.1) is a powerful method for separating out the relaxation and scattering contributions to the measured signal. The velocity of sound is much less sensitive to relaxation effects, provided $\alpha_r \ll k'$, the real part of the wavenumber (see §2.3). An initial attempt at fitting the scattering model (§5.1.5) should first be made to the velocity data, to produce a refined model. Any disparity between this refined model and its fit to the attenuation data may then be ascribed to relaxation or other effects, which can be compared with data for the attenuation spectrum in the continuous phase. Finally, the contribution to the attenuation spectrum from relaxation effects determined earlier may be subtracted from the measured total-attenuation spectrum.

5.1.5 ULTRASONIC METHODS OF PARTICLE SIZING

5.1.5.1 Simultaneous Measurement of Velocity and Attenuation

It is straightforward to measure both velocity (phase) and attenuation (amplitude) simultaneously in modern ultrasound instrumentation. In general both continuous and pulsed methods can be used to do this, and this can often be done in a single apparatus (see §5.1.5.4 and §5.2.3). Examination of Figures 4.6 and 4.7 suggests that measurement of both velocity and attenuation together will give a far more accurate particle size and particle size distribution than either on its own. In general terms, the so called "inverse problem," finding a particle size distribution from the scattering coefficients, is far from trivial. In the case of ultrasound data, solution of the inverse problem is in its infancy. A range of methods for solving the inverse problem are described by Alba (1992). Inversion techniques in the context of light-scattering are discussed by Yan and Clarke (1989). The acoustic case is both more complicated and potentially much more accurate than light-scattering and than acoustic-attenuation-only (Alba, 1992) techniques, because of the availability of both amplitude and highly accurate phase information over many frequency decades.

5.1.5.2 Determination of Particle Size from Velocity and Attenuation

In the "model-independent" approach to the determination of particle size, scattering theory equations for velocity (Equation 4.66) and attenuation (Equation

4.71), the physical data required to solve the scattering theory equations (Table 4.7), and the experimental data in the form of ultrasound spectra (velocity and attenuation as function of frequency) are combined together to obtain a particle size distribution ($\phi_i(r_i)$). It is necessary to constrain $\phi_i(r_i)$ in some way in order to obtain a unique solution (Yan and Clarke, 1989). Model independent methods are computationally intensive, and are very sensitive to errors in the physical parameters required by scattering theory, as well as to errors in experimental data.

The "model-dependent" approach is much less demanding computationally and makes fewer demands on the data. It involves specifying in advance the mathematical form of the particle-size distribution. This approach has the serious disadvantage that an apparently good fit may be obtained with a model distribution which bears little relation to the real one. However, because of its relative ease of implementation, this approach is widely used.

In practice, a compromise between model-independent and model-dependent approaches may be used. The use of both velocity and attenuation data makes this quite feasible in the case of ultrasound scattering. In a given situation, the velocity spectrum (Figure 4.6) and the attenuation spectrum (Figure 4.7, see also Figure 4.12 for bubbles) have quite different forms and respond in different ways to changes in particle-size distribution. These properties of the spectrum may be exploited to devise efficient and accurate methods for the determination of particle-size distributions.

The first step in analyzing the data using the model-dependent approach is to produce a scattering model using the explicit solutions summarized in the previous chapter (Tables 4.4 and 4.5, §4.3.8), modified to account for particle-size distribution (§4.3.10). Data for this model will be more or less complete and missing data must be guessed. The model is then fitted to the velocity spectrum in the long-wavelength limit, using least-squares fitting techniques. The particle-size distribution is initially assumed to be log normal

$$\phi_i(r_i) = \frac{\phi}{\bar{r} \ln \sigma_{\bar{r}} \sqrt{2\pi}} \exp\left[\frac{\ln^2 \sigma_{\bar{r}}}{2}\right] \exp\left[\frac{(\ln r_i - \ln \bar{r})^2}{2\ln^2 \sigma_{\bar{r}}}\right]. \qquad (5.2)$$

where $\phi_i(r_i)$ is the volume fraction of particles of radius r_i; $\bar{r}$ is the mean particle radius; and $\sigma_{\bar{r}}$ is the standard deviation in the droplet radius.

The problem of missing data may be overcome if a series of spectra are available, measured against concentration in the long-wavelength limit. Equation 4.56 may then be used to determine, by experiment, the zero- and first-order scattering coefficients. This eliminates uncertainty as to the values of the physical constants otherwise required to calculate these coefficients. All that is required, then, is to vary $\bar{r}$ and $\sigma_{\bar{r}}$ in order to minimize the sum-of-the-squares difference between the measured velocity and the velocity predicted by the model. It may be necessary to refine the model distribution in an iterative manner, successively applying refined distributions to the velocity spectrum and then to the attenuation spectrum. Alternative model particle-size distributions may need to be explored. An example of the determination of a particle-size distribution from experimental data is shown in section 5.2.2. Strategies for determining particle-size distribution solely from attenuation data are discussed by Alba (1992) and Riebel and Löffler (1989).

5.1.5.3 Bandwidth and Signal-to-Noise Ratio

The bandwidth of ultrasound apparatus is primarily a function of the transducers. Plastic transducers such as those manufactured from PVDF offer the best performance and bandwidth, but are not being manufactured in large numbers at the frequencies required for particle sizing. Plastic transducers also require a very high impedance amplification stage which increases costs, and are generally much more expensive than ceramic transducers. Plastic transducers also require burst radio-frequency (rf) excitation (which employs function generators such as the Hewlett-Packard 4115A, Hewlett-Packard, USA), which is more expensive to provide than the electrical spike generation often used for ceramic transducers. In addition, burst rf excitation which contains a single primary frequency requires that the frequency be changed gradually in order that measurements may be made at each frequency. This is much slower than generating all the frequencies in a single pulse and then extracting the full spectrum using fast Fourier transform techniques. There is no doubt that these technical problems will be overcome as ultrasound particle sizing becomes more widespread.

At present, low-cost ultrasound particle sizing systems are based on ceramic transducers which have a very "far from flat" response and a restricted bandwidth, from 30 kHz to 50 MHz. This necessitates the use of calibration techniques, using a distilled water sample. Nevertheless, very good results have been obtained with these transducers. The higher frequencies can be reached with relatively cheap quartz crystal transducers, although these normally must be assembled by the user.

A variety of techniques can be used to improve the signal-to-noise ratio:

1. Increase the output of the transmitter by (a) tuning the output if the transducer has a high or medium Q, for example by using tuning stubs (GE, USA) at frequencies of above about 15 MHz; (b) increasing the output power of the transmitter; (c) improving the impedance matching of the electrical circuit to the transducer and the transducer to the sample; and (d) improving the coupling of the transducer to the sample.

2. Vary the path length to optimize signal-to-noise (inversely proportional to acoustic path length) and the accuracy of the attenuation or velocity measurement (proportional to path length).

3. Increase the gain of the receiver circuitry. Use digital averaging techniques on repetitive pulses. These techniques are available automatically in oscilloscopes such as those manufactured by LeCroy.

4. Use both pulsed (good for fast velocity measurements) and continuous-wave techniques (good for attenuation and high-accuracy velocity measurements)

5. Ensure that the sample is properly degassed.

5.1.5.4 A Particle Sizing Apparatus—Pulsed Method

The frequency scanning pulse-echo reflectometer (FSUPER) used for ultrasound particle sizing at Leeds (McClements *et al.*, 1990c; McClements and Fairley, 1991, 1992) is illustrated in Figure 5.1.

Key features in this equipment include the machining of the sample cell to optical levels of parallelism and careful thermostating of the sample cell. All the requirements of successful ultrasound measurement described in Chapter 2 apply in full measure to ultrasound particle size equipment. The fast Fourier transform, amplitude, and phase of the signal are all computed by the LeCroy oscilloscope under computer control. The results are then transferred to the computer for analysis using a spreadsheet macro. The diffraction corrections (Khimunin, 1972, 1975) are calculated using MathCad and are integrated into the spreadsheet, so automating the calculation of velocity and attenuation as a function of frequency. In the case of McClements *et al.* (1990c), the particle size was determined by fitting the velocity $-f^{0.5}r$ curve to the experimental data, using particle size as the adjustable parameter.

5.1.5.5 Continuous-Wave Interferometer

Very accurate measurements of phase and amplitude employ interferometric techniques. Interferometry involves combining the received and transmitted signal so that the two may "interfere" with one another. The result is a third waveform whose amplitude and phase depend accurately on the *difference* in these quantities

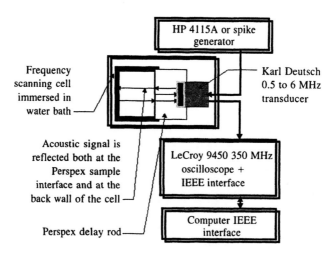

FIGURE 5.1 Frequency-scanning pulse-echo reflectometer.

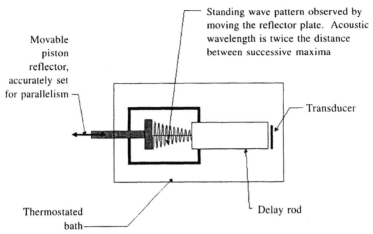

FIGURE 5.2 Schematic diagram of an ultrasonic interferometer.

between the transmitted and received signals. The overall accuracy depends on the phase and amplitude stability of the exciting signal and on the accuracy of calibration. These techniques put very great demands on the engineering of the acoustic cells. Optical levels of parallelism and flatness are necessary, particularly at the higher frequencies. The electronic system may be the same as in Figure 5.1 with the signal generator operating in continuous-wave mode. Greater demands are put on the electronic matching than in the case of pulsed methods, and some form of impedance matching is normally required. Measurements take longer because data must be taken at many settings of the piston reflector, although refinements using piezoelectric pushers may help automate interferometric methods. The interested reader is referred to Edmonds (1981) and Wright and Campbell (1977) for detailed descriptions of ultrasonic interferometers. A more contemporary reference to interferometry can be found in Kaatze et al. (1993), who describe a pulse modification of the continuous-wave interferometer which they claim operates much more quickly. A review of interferometric methods is given in Eggers and Kaatze (1996).

A schematic diagram of an ultrasonic interferometer appears in Figure 5.2. The comments on temperature control in Chapter 2 are very relevant here, particularly because it is necessary to maintain a constant temperature over extended periods. Poor temperature control can completely nullify the accuracy of instruments of this sort. Temperature gradients can have a similarly disruptive effect on experiments. This is why the diagram indicates that the whole ultrasound cell is maintained in a temperature-controlled bath.

5.1.5.6 Commercial Particle Sizing Apparatus

A number of ultrasonic particle sizing apparatuses are now available or are near to being marketed. In Table 5.1 manufacturers of ultrasound particle sizing apparatus are listed.

TABLE 5.1   Manufacturers of Ultrasound Particle Sizing Apparatus

Manufacturer	Equipment name	Operating principle
Sympatech, Germany	Opus	ultrasonic forward scattering
Penkem, USA		ultrasonic forward scattering/ electroacoustics
Malvern, UK	Ultraspec	ultrasonic forward scattering
Matec, USA	Zetasizer	electroacoustics

5.1.5.7 Electroacoustics

An alternating electric field, when applied to an emulsion or suspension containing charged particles, will cause the particles to oscillate with respect to the fluid continuum. The amplitude of the oscillation will depend on the charge on the particle, on the viscosity of the continuous phase, and on the inertial difference between the particle and its surroundings. The analysis of the problem follows the lines developed earlier for the viscoinertial effect, but where it is the electric field acting on the particle which is forcing the motion rather than an impressed sound field. It is also possible to operate with an input sound field and an output electric field. In the first case an electrode is the transmitter and a piezoelectric transducer the detector, and in the second case the roles of electrode and piezoelectric transducer are reversed. It should be noted that the scattered acoustic field in the first case and the scattered electric field in the second are strongest along the z-axis, i.e., parallel to the exciting field.

The use of this phenomenon for particle sizing and the determination of zeta potential has been pioneered by O'Brien and his group (O'Brien, 1988, 1990; Wade *et al.*, 1995; O'Brien *et al.*, 1994, 1995), whose research has been incorporated into the Acoustosizer (Matec Applied Sciences, USA). The cell consists of a sample space (volume approximately 400 ml) between two parallel-plate electrodes bonded to glass delay rods (Figure 5.3). An alternating voltage pulse is applied across the electrodes, and the sound waves generated by the particle motion are detecting by a transducer bonded to the rear face of one of the delay rods. The measurement cell comprises pH, conductivity, and temperature probes and a stirrer. The entire system is automated.

A detailed account of the theory of electroacoustics can be found in O'Brien *et al.* (1995). Although the theory proceeds along the same lines as those indicated in this chapter, it is complicated by the presence of the electric field and its coupling to the particle. On the other hand, thermal effects are reduced to the second order because the "first-order" sound field is present only as the scattered dipolar field from the particle. The advantage of the technique is that the dynamic mobility of the particle may be obtained from the electrosonic analysis signal. ESA must take into account the varying phase between the particle and the applied electric field, and the limits defined in Table 4.5 apply. The Acoustosizer uses ESA measurements to obtain a dynamic mobility spectrum between 300 kHz and 11.5 MHz. The

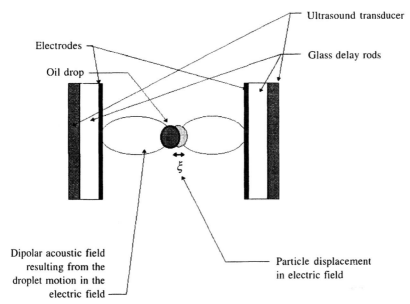

FIGURE 5.3 The electroacoustic effect.

particle averaged dynamic mobility $\langle \mu_d \rangle$ is defined by O'Brien as the complex quantity whose magnitude is equal to the velocity amplitude per unit field and a phase equal to $\omega \Delta t$, where Δt is the time delay between the field and the velocity an ω is the frequency of the applied electric field.

$$V = C\phi \frac{\Delta \rho}{\rho_1} \langle \mu_d \rangle Z, \qquad (5.3)$$

where V is the voltage detected by the acoustic transducer, C is an instrument constant, and Z is a second instrument constant which depends on the acoustic impedances of the suspension and the glass delay rods. $\Delta \rho$ is the density difference between the particles and the solvent and ρ_1 is the density of the solvent.

A number of assumptions need to be made in the determination of the dynamic mobility, and the reader is referred to O'Brien et al. (1995) for a more detailed discussion. The zeta potential (electrokinetic potential) may then be obtained using the formula

$$\mu_d = \frac{\varepsilon \zeta}{\eta_s} f\left(\frac{\omega r^2}{v}\right), \qquad (5.4)$$

where ε is the permittivity of the solvent, ζ is the zeta potential, v is the kinematic viscosity ($= \eta_s/\rho$), and f is a complex function of $(\omega r^2/v)$, where r is particle size. Particle sizing assumes that there is a log-normal size distribution and proceeds by fitting the measured dynamic mobility spectrum to a theoretical spectrum obtained from Equation 5.4. The lower limit of particle size measurement by the Acoustosizer is 0.1 μm, defined by the operating frequency range of the instrument and the lack of a frequency-dependent response in the long wavelength region defined by the limits in Table 4.5 (see also Figure 4.8). The instrument takes into account particle–particle interactions in a semiempirical way, based on the results of measurements in concentrated systems.

5.1.5.8 The Future—Measurement Systems

In future we may expect to see ultrasound instruments alongside Light-scattering apparatus for particle sizing. Since these instruments are complementary, instrument manufacturers may integrate the various sizing techniques into a single instrument. NMR may be integrated with an ultrasound measurement, too. In the area of process measurement, ultrasound size measurement may become very widely used because of the relative ease with which it can be integrated into process control and measurement systems. Ultrasound size measurement apparatus will also carry out concentration measurements and are even capable of integrating with ultrasound Doppler flow meters. The result of this would be a single instrument and a single transducer capable of measuring flow rate, flow profile, particle size, and particle size distribution and concentration. The technical means to do this exist now. Integration of ultrasound concentration measurement with density measuring devices will give, in addition to all the information previously mentioned, data on compressibility, which in turn could give process information on the molecular state of solutes. Finally, the advent of more powerful computers, allied to sophisticated electronic data acquisition systems, means that rapid development in the area of acoustical particle sizing is likely.

5.2 PROPAGATION IN VISCOELASTIC MATERIALS

5.2.1 INTRODUCTION

Viscoelastic materials are generally characterized by a zero or small apparent yield stress over infinite time scales and by non-Newtonian behavior in the case of *viscoplastic* materials (Ross-Murphy, 1995). The condition on yield stress must be applied because any solid material can be made to exhibit viscoelastic behavior if the stress conditions are extreme enough. One only has to look at cold die pressing of metals to see strong, hard materials flowing like liquids below their melting point. All liquids will behave like a solid at high enough strain rates. Some materials that are liquid over long time periods can behave like solids when stressed over short time periods (shear thickening) and vice versa (shear thinning). Examples of shear thinning materials include toothpaste, set yogurt, gravy, and some paints. A good example of a shear thickening material is custard powder mixed with a little sugar and milk, which behaves like a liquid when stirred slowly but thickens into a solid when it is stirred more quickly. The time period over which a stress is to be applied is relevant here. Glass behaves like a liquid over centuries, yet it is clearly a solid material for most practical purposes.

5.2 Propagation in Viscoelastic Materials

From the point of view of this text, the question is to what degree can the scattering theory developed in Chapter 4 for fluid or solid particles suspended in a fluid medium be applied to viscoelastic materials? The general approach to this question has been outlined in §4.3.3.2 and may be summarized as follows.

The stress tensor ($\mathbf{P}$) must be written down in terms of the strain-rate tensor ($\dot{\mathbf{S}}'$) and the functional dependence of stress upon strain rate may be viewed as a generalized viscosity:

$$\mathbf{P} = \eta(\dot{\mathbf{S}}'), \qquad (5.5)$$

where this relationship is defined in Equation 4.15. Written in this way, the viscosity cannot be defined as one or two numbers such as the bulk or shear viscosity, but is instead a function which defines the relationship between two tensors. It is only when the stress and strain rate have simple forms that we can write *viscosity = stress/strain-rate*. The various forms of the viscosity are then derived in terms of the Lamé constants by defining the time dependence of the bulk and rigidity moduli (Equation 4.16 for a solid and Equation 4.19 for a liquid). The appropriate stress tensor may then be used to derive the dispersion relations for the systems, appropriate boundary conditions are applied, and the scattering coefficients derived. Finally, the propagation equations may be derived by inserting simplified forms of the scattering coefficients, under the appropriate conditions, into the dispersion relation, Equation 4.22. This is described in more detail in the appendix, where the main results for propagation in viscoplastic materials are derived.

In weakly aggregated systems, the basic single-particle scattering will remain largely unaffected. It is the multiple scattering effects which will be sensitive to the state of aggregation, as has already been indicated in the previous section (see also §5.1.4). A simpler approach than that just described, which can be derived from scattering theory (see Appendix, Equation A.7 for solids and Equation A.19 for viscoplastic materials) under the correct limiting conditions, is to use a version of the Wood equation applicable to solids:

$$v = \sqrt{\frac{B + \tfrac{4}{3}G}{\rho}} = \sqrt{\frac{M}{\rho}}, \qquad (5.6)$$

where B is the bulk modulus, G, the rigidity modulus, and M, the elastic modulus of the material. The rigidity modulus G is on the order of 10^9 times smaller than the bulk modulus in the case of weak gels such as set yogurt. It is not then surprising that efforts to detect changes in the storage modulus (real part of the elastic modulus) of weak gels (Audebrand *et al.*, 1995) have been unsuccessful. Changes observed by Audebrand *et al.* (1995) in attenuation in weak gels during the gelling process are almost certainly associated with changes in the state of aggregation of the monomer, rather than with the appearance of a shear modulus.

Javanaud and Robins (1993) have suggested that the appearance of a network associated with gelation may permit the appearance of a second acoustic mode which is predicted by the coupled phase theory of Biot (1956a, b; Ogushwitz, 1985a, b), and modifying the normal mode of propagation. There is no doubt that the Biot theory merits serious consideration. The Harker and Temple (1988) approach, adopted by Evans and Attenborough (1996), may offer a way forward if it can be reconciled with scattering theory's detailed microscopic model. However, there are a number of ways in which single and multiple particle scattering may be modified by aggregation, and these effects will be in addition to any predicted by Biot theory.

The processes that lead to changes in ultrasound velocity during droplet flocculation will also take place in micelle aggregation. In the case of the aggregation of casein submicelles (Chu *et al.*, 1995), the forces involved are stronger than the reversible flocculation discussed earlier. It may therefore be assumed that the casein micelle behaves as a single-particle, rather than a floc. If this is so, then the effects of casein micelles, casein submicelles, and Tween 20 micelles can be calculated using scattering theory in the long wavelength limit (Povey, 1996). The results of these calculations appear in Figures 5.4 and 5.5.

The predicted attenuation in Figure 5.4 at 1 MHz for Tween and casein micelles is too small to be measured by most pulse methods, but by 10 MHz the effects are large and it should, in principle, be possible to monitor the state of aggregation of casein micelles at this frequency. The same applies to the predicted changes in ultrasound velocity shown in Figure 5.5. It is clear from the figures that apparatus with a bandwidth up to 100 MHz will have a very good chance of effectively monitoring aggregation in these systems.

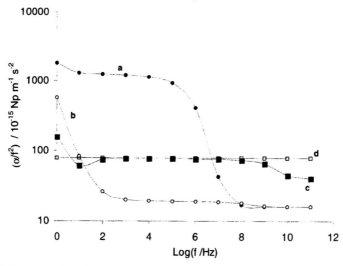

FIGURE 5.4 Predicted frequency-dependent attenuation from (a) casein micelles (50 vol%, corresponding to 12 wt%, 0.2 μm), (b) casein submicelles (100 nm, Borthakur and Zana, 1987), and (c) Tween 20 micelles (2 wt% Tween 20 in water, 4 nm). Some data for the calculation has been abstracted from Griffin and Griffin (1990).

5.2 Propagation in Viscoelastic Materials

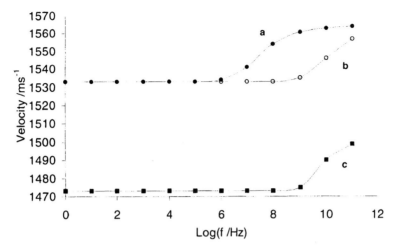

FIGURE 5.5 Prediction of the frequency-dependent velocity for the systems plotted in Figure 5.4. Annotation is the same as for Figure 5.4.

The results of experiments performed on the basis of the preceding predictions are very interesting. Measurements over the range 1 to 115 MHz, carried out with the Malvern Ultraspec using a bimodal, log-normal fit, indicated that a bimodal distribution of casein micelles (4 wt% sodium caseinate dissolved in water at 25°C) existed (Figure 5.6), with a main peak at 21 nm and a much smaller peak around 530 nm. The fit, however, was very poor with a SSD of 17%, which would

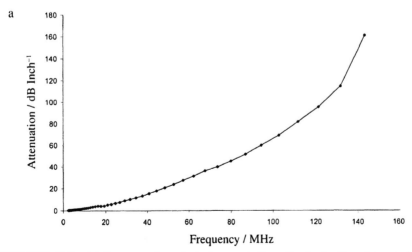

FIGURE 5.6 (a) Frequency dependent ultrasound attenuation in 4 wt% sodium caseinate dissolved in water at 25°C. (b) Particle size distribution computed from the attenuation plot of (a).

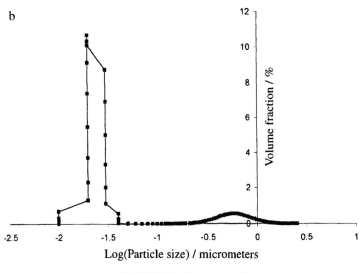

FIGURE 5.6 *(continued)*

normally be considered unacceptable. Little change was observed when calcium ions were added to the solution, which are expected to aggregate the micelles. Photon correlation spectroscopy was carried out in parallel (on samples extracted from the Ultraspec tank and then diluted), confirming the existence of the two peaks at 21 nm and 530 nm but indicating that the volume fraction of material at 530 nm was very small indeed. While the PCS peak at 21 nm was reduced in size after the addition of calcium, little change was seen in the 530 nm peak. It seems, at this early stage, that the aggregates, which are definitely created by the addition of calcium, have a very different scattering profile to the 21 nm micelles. It is probable that this is the cause of the very poor fit of the model to the ultrasonic data. It is very encouraging to see that the technique clearly identified the submicelles, as predicted. However, these preliminary experiments suggest that our initial hypothesis, that casein submicelles aggregate to form micelles which have similar acoustic properties to the submicelle, is wrong.

5.2.2 MEASURING AGGREGATION IN VISCOELASTIC MATERIALS

5.2.2.1 Introduction

There is considerable amount of experimental evidence for the ultrasound detection of aggregation in a range of systems. These include ordinary milk (Bachman *et al.*, 1978; Griffin and Griffin, 1990; Miles *et al.*, 1990) skimmed milk (Griffin and Griffin, 1990; Benguigui *et al.*, 1994; Gunasekaran and Ay, 1994), blood platelets (Horak, 1972; Mahony, 1987; Mahony *et al.*, 1989;

5.2 Propagation in Viscoelastic Materials

Kitamura *et al.*, 1995; Chabance *et al.*, 1996), asphaltene (Bouts *et al.*, 1995; de Boer *et al.*, 1995); cyanazine slurry (Holmes and Challis, 1993) and *n*-hexadecane oil-in-water (McClements, 1994).

5.2.2.2 Detecting Aggregation with Ultrasound Profiling

In §3.5, the use of the ultrasound profiling technique was discussed. This is a very powerful method for studying aggregation and particle size in emulsions in its own right. All that is necessary is to follow the changes in concentration of the dispersed phase as a function of time and height to learn a great deal about the changes taking place in an emulsion (Pinfield *et al.*, 1994).

In Figure 5.7 are plotted the group velocity for a nominal frequency of 2.25 MHz and attenuation for a casein-containing emulsion like that shown in Figure

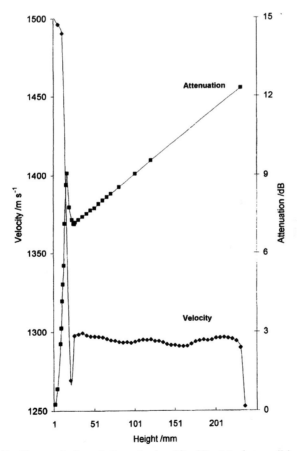

FIGURE 5.7 Group velocity and attenuation in a 10 vol% *n*-tetradecane oil-in-water emulsion containing 4 wt% sodium caseinate, 30°C, pH 6.8, d_{32} = 0.5 μm. The profile was taken after three hours. The attenuation plot was obtained from the data in Figure 5.8 and is the integrated attenuation relative to distilled water (Povey, 1996).

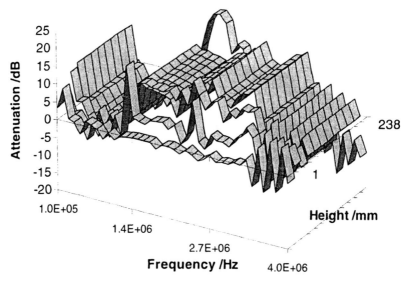

FIGURE 5.8 Ultrasonic signal amplitude, relative to distilled water, plotted against frequency and height in the profiler tube (Povey, 1996).

3.16, but at the much lower oil concentration of 10 vol%. These emulsions creamed quickly but then formed a "stable" emulsion phase consistent with a gel. Examination of Figure 5.7 shows the presence of a very thin oil-rich layer below 51 mm (the height steps at the bottom of the sample tube have been expanded to highlight effects in the serum in both Figures 5.7 and 5.8). This could be seen by eye, and it seems that very fine, unflocculated oil particles which have been left behind in the serum phase by the larger oil droplets and flocs, finally do begin to cream and then find their progress hindered by the gelled emulsion. The boundary between the serum and the emulsion then becomes the gathering point for the oil-rich layer, which can be seen clearly in both the velocity plot and the attenuation plot. This effect is made more marked by the tendency of the back flow around the larger upwardly mobile particles to cause the smaller particles to move downwards.

However, there is still a great deal that is missing from the picture, and quantitative or semiquantitative information about particle size and the state of aggregation of the systems is very valuable.

5.2.2.3 Computer Modeling

A lot can be learned by comparing experimental results with computer models. For example, Pinfield *et al.* (1996) have modeled creaming and flocculation in emulsions using a lattice model. The model simulates the simultaneous creaming and flocculation of emulsions. A cubic lattice model is used, with particles bonding together and breaking apart according to predefined probabilities. Creaming and diffusion are included, with different rates for different sizes of floc. Results are presented in the form of concentration profiles as a function of height, and as a fractal dimension variation of the particle networks formed. In the numerical simulations, an increase in creaming rate was observed before a particle network was formed. These results explain some features observed experimentally in creaming and flocculating emulsions, but do not reproduce the rapid serum development seen in some cases. However, modelling can be valuable even if the experimental data are not accurately reproduced. This is because the construction of the model forces a careful quantitative examination of all the processes likely to be of importance. A discussion of computer simulation as a tool for studying emulsions and aggregation processes can be found in Dickinson and McClements (1995a).

5.2.2.4 Aggregation of Casein

Equations A.19 and A.20 give the storage (M') and loss (M'') moduli in terms of velocity and attenuation. An example of such calculations for casein micelles when aggregation is induced by addition of acid or rennet is given in Table 5.2.

To appreciate the results in Table 5.2 it is necessary to understand a little about the aggregation of casein micelles. Casein is a complex protein obtained from milk which comprises three forms, α-casein, β-casein, and κ-casein. Together the three types form a micelle in which the hydrophobic parts of the protein tend to group together inside while the more hydrophilic parts come in contact with water. The micelles can be induced to come together to form aggregates through the effects of enzymes, pH, or heat. This forms the basis of processes such as cheese making and yogurt manufacture. Casein is also widely used as a stabilizer.

The process of aggregation often gives rise to gelation, and this has been achieved in the experiments shown in Table 5.2 by either the addition of acid or the use of an enzyme, rennet. Both processes result in gelation of the casein. There is very little difference between acid gelation and enzyme gelation, so far as the final strength of the gel is concerned. The results at all frequencies indicate that the difference between the storage moduli for the renneted and untreated milk is insignificant. This is consistent with the earlier observation that the contribution of the rigidity to the total modulus in systems such as these is very small. The larger difference in the storage modulus in the case of acid-treated milk is probably due to disruption of the core of the casein micelle, changing the compressibility of the dispersed phase. The changes in velocity reported by on rennet and acid gelation of skimmed milk by Benguigui *et al.* (1994) (Table 5.2) can be explained partly by the changes in the state of aggregation of the casein micelles, which will certainly have a significant effect at the 60 MHz employed by these workers (see Figures 5.4 and 5.5). The much larger effect observed from acid gelation (Table 5.2) can be explained by changes in the compressibility of casein arising from the disruption of the core of the micelle caused during acid gelation (Povey, 1996). Rennet, on the other hand, has little effect on the micelle core (Benguigui *et al.*, 1994), and its effects on ultrasound propagation are likely to be much smaller.

The moduli in Table 5.2 are clearly very frequency-dependent, but we are far from having a model of the elastic properties of gels which can explain the frequency dependence we see. It is likely that the high-frequency ultrasound properties are responding to aggregation processes which precede gelation. We have already seen that the scattering will depend on the distribution of particles as well as their overall concentration. So aggregation may be expected to change the scattering simply by virtue of the changing distribution of particles. It is also possible that aggregation may affect the multiple scattering by modifying the diffusion of particles. This is because the Brownian motion of particles undergoing diffusion will create a Doppler shift in the scattered signal. Once part of an aggregate, this motion will be considerably reduced, the Doppler shift will decrease and the scattering will be modified. Although other processes may be at work, such as overlap of thermal and shear waves, and the appearance of interparticle forces, it is not actually necessary to invoke these to see that aggregation can change the acoustic properties of materials.

TABLE 5.2 Experimental Results at Frequencies between 0.1 Hz and 60 MHz for Aggregation of Casein Micelles[a]

Conditions	Variable	Results		
		Ungelled	With acid	With rennet
0.1 Hz, 25°C (Benguigui *et al.*, 1994)	G' /Pa			4.20
0.1 Hz, 25°C (Benguigui *et al.*, 1994)	G'' /Pa			2.16
1 Hz, 25°C (Griffin and Griffin, 1990)	G'' /Pa			25–250
2 MHz, 30°C (Cosgrove, O'Donnell and Povey, 1996)	Δv /ms $-$ 1		~0.5	0
2 MHz, 30°C (Cosgrove, O'Donnell and Povey, 1996)	M'' /Pa	0.4×10^{-6}	variable	0.4×10^{-6}
60MHz (Benguigui *et al.*, 1994)	Δv /ms $-$ 1		3.5 44°C	0.2 25°C
60MHz (Benguigui *et al.*, 1994)	M' /Pa	2.481×10^9	2.492×10^9 44°C	2.482×10^9 25°C
60MHz (Benguigui *et al.*, 1994)	$\Delta M''$ /Pa		20.3×10^6 44°C	0.2×10^6 25°C

[a] From Povey (1996).

The discussion thus far suggests that there are at least two processes involved in gelation, so far as ultrasound measurement is concerned. The first, and the most important from the ultrasound point of view, is the detection in the change of the state of aggregation of the particles, be they oil droplets, blood platelets, or casein micelles. The second process follows from the first: The aggregates continue to grow until they fill the sample space and acquire a yield stress. This process is hardly detectable with compression ultrasound.

In conclusion, there is no doubt that ultrasound can be used to detect aggregation. However, this is not a straightforward particle size effect, but is probably a second order effect arising through changes in multiple scattering. This will be related to the size of the particles making up the aggregate. It is only in the case of "hard" aggregation, that a straightforward particle size effect is likely to be seen. Ultrasound spectrometry has a great deal to contribute to the study of aggregation processes, but the mechanisms by which aggregation affects ultrasound propagation are poorly understood. Consequently, the technique is semiquantitative. Integration of ultrasound spectrometry with other techniques such as ultrasound profiling and NMR may produce great benefits.

5.2.3 FREQUENCY-DEPENDENT ULTRASOUND PROFILING

With careful attention to the acoustics, it is possible to measure frequency-dependent velocity and attenuation in the Leeds ultrasound profiler. Data is presently available between 0.1 and 10 MHz and specimen results are shown in Figure 5.8.

The attenuation data in Figure 5.7 suggest that there is considerable variation between the bottom and the top of the serum. The velocity data suggest that this is not due to variation in the oil volume fraction, so it may be that a polymer concentration effect is observed here, arising from the displacement of polymer from the flocculated emulsion. At the very top of the tube, the velocity data indicate that there is the beginnings of a cream, whose development has presumably been arrested by the gelation of the emulsion. The variation in the attenuation from the bottom to the top of the emulsion is hard to explain, but may be related to the displacement of polymer during the aggregation process and its subsequent extrusion from the gel during its slow collapse under the influence of gravity. This process is sometimes called "syneresis."

The attenuation data in Figure 5.7 were computed from the Fourier transformed ultrasonic signal whose frequency space representation appears in Figure 5.8. The serum region at the bottom has been plotted at 1 mm intervals up to 51 mm in order to highlight the changes that have taken place in the serum. Visually, the serum appears transparent and uninteresting. However, there appears to be a great deal of frequency-dependent scattering occuring within the serum layer. This suggests a considerable heterogeneity in this part of the sample. The difference in behavior between the serum and the emulsion is dramatic. It may be helpful to consider an analogy with Light-scattering at this point. If 0.1-MHz ultrasound corresponds to the infrared and 4-MHz ultrasound to ultraviolet, a hypothetical false-color picture of Figure 5.8 could be constructed. In this picture, the acoustic signal from the serum would correspond to a sample which appears opalescent in white light, showing different colors in different parts of the sample. In this analogy, the emulsion would appear more or less uniformly "grey." The incipient cream layer would appear blue as the lower frequencies were blocked out and the higher ones transmitted, possibly due to Rayleigh scattering of the signal.

We are now in a position to put the detection of flocculation and aggregation into the context of acoustic scattering generally. Both the frequency-dependent velocity and the attenuation indicate that the scattering of ultrasound decreases appreciably when depletion flocculation takes place. This has been explained on the basis of the overlap of the thermal and shear waves between adjacent droplets (McClements, 1994). At 3 MHz and 20°C the thermal and viscous decay lengths are 120 and 500 nm, respectively. The separation of oil droplets in a floc must be less than 10 nm for flocculation to occur. This is the approximate diameter of the Tween 20 micelles used in the preparation of the emulsions in these experiments. However, the changes in ultrasound velocity occurring as a result of flocculation are very small. In the experiments reported in McClements (1994), the velocity change was just 2 m s^{-1}. Experience with ultrasound profiling supports the hypothesis that velocity changes associated with flocculation are small. If this were not the case, then the technique would not work because it would not be possible to convert velocities to concentrations through the use of the modified Urick equation. So attenuation changes are likely to be more useful in detecting aggregation. Finally, the profile of Figure 5.8 indicates that there is a wealth of detail contained in the ultrasound data, waiting to be unlocked by further theoretical and experimental developments.

5.2.4 PARTICLE SIZE EFFECTS IN ULTRASOUND PROFILING

The velocity to concentration conversion outlined in §3.7 depends, among other things, on the constancy of the particle size distribution throughout the creaming process. Computer modeling (Pinfield et al., 1994) indicates that the particle size distribution can change dramatically in polydisperse samples. There are a number of reasons for this, some of which have been mentioned already. For one reason or another a fractionation process takes place among the oil droplets in the emulsion, which reaches its most developed form in the cream. This manifests itself in the ultrasound signal, in the form of rapidly changing velocity and attenuation, and this is sometimes observed in the cream. Unfortunately, the renormalization process can give a false sense of security because the total oil fraction can remain equal to the original oil fraction, while the apparent volume fraction, as determined from the velocity, fluctuates wildly. This condition is normally associated with greatly increased attenuation, so it is monitored automatically by software and the condition flagged in the data. It is also possible that flocculation removes some of the particle size variation by incorporation of a range of particle sizes into a single floc.

5.3 BUBBLES AND FOAMS

Very many fluids contain gas; sometimes the bulk of the fluid is gas, as is the case with foams. Examples of such systems include beer foam, ice cream, and foams of all kinds which abound and are often unwelcome in processing. In the case of fluids, a gas dispersion is unstable, and the principal destabilizing mechanisms are as follows (Fairley *et al.*, 1991; Fairley, 1992): (a) creaming due to the large density difference between gas and liquid; (b) foam drainage and subsequent collapse due to high interfacial tension; and (c) gas migration from the smallest bubbles into larger bubbles due to the higher Laplace pressure within the smaller bubbles, via diffusion of gas through the continuous phase. This last mechanism is called disproportionation and is an extreme form of Ostwald ripening. It is often desirable to know the bubble size, size distribution, and phase volume of gas. However, these are not so easy to measure. Gravimetry is insensitive and often hard to apply. Microscopy is hampered by difficulties in representative sampling and by the wide range of bubble sizes encountered in real systems. So there is a need for a technique which can characterize bubbly systems in real time and without special preparation. The great sensitivity of ultrasound to bubbles strongly suggest its use for measuring bubbly systems (§4.4); however, it is also clear from the theory of bubble resonance that measurements in concentrated systems are likely to be of an empirical or semiempirical nature. Nevertheless, reflection techniques such as those described in §5.1.5.4 can give rapid and repeatable readings (see also Manetou, 1990).

In Figure 5.9 the real and imaginary parts of the reflection coefficient are plotted for unwhipped and whipped full fat set yogurt, using the pulse-echo reflectometer (§5.1.5.4). The effect of the incorporation of gas bubbles is very clear. The measurements were very repeatable and qualitative differences could clearly be seen. Such measurements were successfully repeated on cream and beer foam. Comparison with theory (§4.4) indicated that, although the behavior of the individual bubble resonance could be seen, it was not possible to predict the effects of concentrated systems of bubbles on ultrasonic propagation. So it seems unlikely that bubble sizing can be carried out at present, except in a qualitative or semiempirical manner.

5.4 AUTOMATION AND COMPUTER TOOLS

The complexity of the interaction between sound and matter is an obstacle to the use of ultrasound for process monitoring. For example, in the case of ultrasound

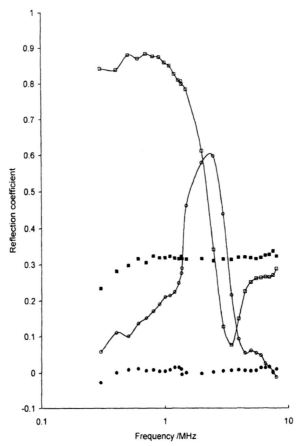

FIGURE 5.9 The measured real part (■, unwhipped; □, whipped) and imaginary part (●, unwhipped; ○, whipped) of the reflection coefficient in full-fat set yogurt (Fairley, 1992).

profiling, it is desirable to account for the effects of particle size distribution, and methods have been presented to help do this (§3.7.4.1). These methods require the use of computer and automation techniques, which can help surmount the difficulties of complex analysis and unlock the information hidden within the sound field. Although bats can see with sound (§1.1), human beings cannot. But with the help of computers, we can begin to see a whole new world unfold before us, revealed to us by the use of sound, which can penetrate materials which are opaque to light.

The purpose here is to guide the reader in the use of computer techniques for the automation of acoustic data analysis and the control of instrumentation. The chosen tools are the personal computer, the IEEE bus, and RS232 serial communications. Very few references are given from here on. In general, the best sources of reference are the instrument manuals themselves, the manuals supplied by the software manufacturers, and their associated help and sample files. It is recommended that the reader acquire the equipment and manuals and learn directly from their use. The catalogs of instrument manufacturers are also a valuable source of information in this rapidly changing area.

Finally, if ultrasound instrumentation is to be widely used, it must be taken out of the province of the ultrasound expert and placed in the hands of scientists and engineers in general. Computers can help achieve this aim, by simplifying operation and encapsulating expert knowledge in easy-to-use instrumentation.

5.4.1 THE COMPUTER AS CONTROLLER

The personal computer is capable of carrying out simultaneously the functions of instrument control, data acquisition, analysis of ultrasound data, and presentation of data to the user. Although the power of personal computers at present is not such that they can quickly calculate scattering matrices like that of Equation 4.27, explicit solutions such as those presented in Table 4.4 are easily calculated. The following discussion assumes that the controller is a personal computer (PC) comprising a 90 MHz Pentium processor with 16 Mbyte of random access memory (RAM) and 1 Gbyte of hard disk storage. The software tools employed are MathCad, Visual Basic (VB), National Instruments GPIB drivers and Excel. The example of an ultrasound profiler will be used to illustrate the role of the computer in ultrasound instrumentation.

In Figure 5.10, the Visual Basic control program acts like a kind of "glue," linking all the elements of the apparatus together into a single coordinated unit. Visual Basic is an *event-driven* programming environment, unlike older algorithmic programming languages such as FORTRAN. "Event-driven" means that different parts of the program code are triggered by events which are generally outside the control of that code. These events may be a mouse click on a text box on the visual display unit, for instance, or a signal from a timer. The result can be a program which is easy to use and which greatly simplifies the complex operations involved in ultrasound profiling.

5.4.2 WINDOWS

Event-driven approaches to programming are appropriate in a multitasking environment such as that provided by the various Windows operating systems. However, it does require thought about what else is taking place within the computer, and the resources that are available to the program at different points in the operation of the apparatus. For example, the Visual Basic program transfers its results to an Excel spreadsheet for further processing. The Excel spreadsheet and its associated macros (programs comprising Excel commands or those of another software tool), and possibly its own Visual Basic code, runs as an independent process, communicating with the control program and with the graphical display interface (GDI). Thus, Excel consumes its own resources (RAM,

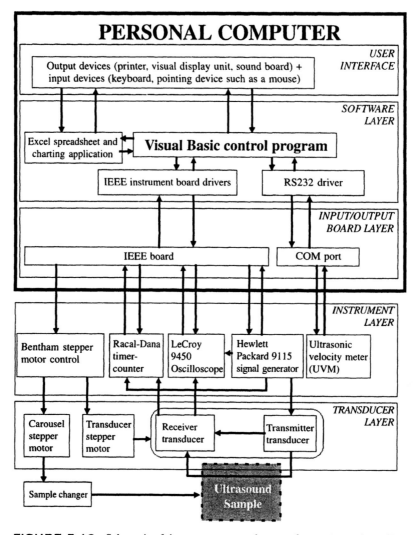

FIGURE 5.10 Schematic of the computer control system for an ultrasonic profiler.

computer processing unit [CPU] time, disk storage, and GDI), which affect the running of the control program. The balancing of these resources is a tricky aspect of Windows programming, which Windows 95 has gone a little way toward addressing, and for which Windows NT is much better suited. The balancing of resources involves *multitasking,* which means the ability of a computer *operating system* (the computer control program) and the computer central processing unit to give the impression that many programs (tasks) are running at once. What is in fact happening is that the operating system allocates control of the CPU and memory among the different programs at a rate which gives the impression that they are all running at once. In fact, only one program is in control at a time, typically for around 20 µs, before control is passed to another process. It is important to know this because earlier versions of Windows ran on the basis of *cooperative multitasking*. In this version of multitasking it was always possible for a program to seize control of the CPU and retain it forever, a major cause of the unreliability of pre–Windows 95 operating systems. When programming for the Windows 3.x environment, it is essential deliberately to write code which releases the CPU for other processes, when appropriate. In other words, code must be "socialized" and every program must be written considering the other processes likely to be running at the same time. Later versions of Windows run *preemptive multitasking*, in which the operating system can always seize control from any processes it initiates. In the case of Windows 95, it is still possible to start processes outside the control of the Operating System. This is a feature which has been included to retain compatibility with earlier versions of Windows and DOS (disk operating system, the operating system which preceded Windows). In the case of Windows NT, it is not possible to circumvent the operating system, making for a more stable system.

Windows 95 and Windows NT are 32-bit operating systems whereas Windows 3.x is 16-bit. The terms 16- or 32-bit refers to the size of the numbers that the CPU can handle. 32-bit operating systems are faster and more accurate than their 16-bit predecessors.

5.4.3 PROTOTYPING

One approach to programming is to break the required task up into its component parts and write the code to implement each task as a separate program. Each separate task is tested and developed independently, before finally being integrated into the growing control program. This has the advantage of delaying the consideration of interactions between different aspects of a program until after the programmer is sure that one piece of code operates satisfactorily as an independent entity. Once the code is tested and works independently, then problems due to interactions between different parts of the code can be isolated and solved.

Testing of the code can occupy more than 90% of the time. This is because program users generally think in a different way from people who conceive the program. "Naïve" users (every user is naïve where a new program is concerned) expose problems and ways of using code which the developers did not anticipate. A program written for other people to use must be straightforward for them to use, or there is little point in programming!

Visual Basic allows very rapid "prototyping" of code, because it handles the details of the GDI, obviating the need to program this aspect of the computer operation. It is also possible to interrupt program execution, and examine and change code while the program is running and controlling the instruments. This is a very useful feature where ultrasound instrumentation is concerned because the code directly controls the instruments, and the effect of code changes on instrument behavior can be tested very quickly.

Programs for use in the laboratory sometimes do not get beyond the prototyping stage, and are under continual development as instrumentation development occurs. The program is connected to instrumentation outside the computer through computer interfaces which adhere to protocols (a set of rules) such as IEEE and RS232C. Computer interfaces themselves are connected within the computer by the computer *bus* (a set of wires which connect the digital components of the computer). The nature of this bus is an important factor in the performance of the computer because the speed with which data is transferred between the components of the computer limits its speed.

5.4.4 RS232C

RS232C is a *protocol* for serial communications between electronic instruments and computers. *Serial communications* involves the transmission of one bit (the elementary unit of data; it can have no more than two values) at a time and is therefore slower than parallel communications such as that using IEEE and printer ports, which can transmit one or more characters at a time. This protocol can be implemented with as few as three wires, which means that it can be very cheap. As a result, it has become the generic computer interface and is found in every PC. The RS232C protocol is loosely defined in practice, and this can cause problems in its implementation. The RS232C protocol defines the role of the wires making up the interconnection; the code associated with the protocol has been adopted by convention, but is not strictly enforced. Special interfaces must be used if more than one instrument is to be coupled to a single RS232C interface.

For various reasons, programming a serial interface can be a much more complicated task than programming IEEE.

5.4.5 IEEE BUS

This is a protocol for electrical connection between a computer and instrumentation which was first developed by Hewlett-Packard for their instruments (called HPIB or Hewlett Packard instrument bus) and was then adopted by the IEEE (Institute of Electrical and Electronic Engineers) as an industrywide standard (the general purpose instrument bus or GPIB). It is a strictly defined protocol (both electrical and code), which results in guaranteed success in interconnection between any instruments manufactured to this standard. It is a parallel interface and therefore is faster than corresponding three- or five-wire systems. Thus IEEE systems generally have a higher bandwidth (maximum rate of data transmission) than corresponding serial systems. However, it is much more expensive than serial communications, resulting in a premium at each node (point of connection). The IEEE bus can accommodate up to 32 instruments in a single system. It is easy to add another instrument to a system. IEEE tends to be used for more sophisticated instrumentation coupled over short distances. It is widely used in laboratory instrumentation.

5.4.6 INSTRUMENT PROGRAMMING

Many instruments which support the IEEE instrument bus can be programmed in their own right. To do this successfully it is necessary for the programmer to understand how to operate the instrument. The instrument manual will detail the *codes* (combinations of characters with defined meanings or relationships to other groups of characters; these are generally three- or four-letter and -number codes in the case of IEEE) required to program the instrument. These codes are transmitted to the instrument from the VB control program. The control program actually communicates with another program, called the driver, which translates the three- and four-letter code combinations into the IEEE bus code and operates the IEEE card in the computer (Figure 5.10). The driver will be supplied by the manufacturer of the IEEE card. In a fully programmable instrument such as the LeCroy 9450 (LeCroy, U.K.) or the Racal-Dana timer-counter (Racal Instruments, U.K.), every aspect of instrument operation can be controlled by the IEEE bus. In fact, it is often possible to do more things over the IEEE bus than it is via the instrument front panel.

Other instruments (such as the Bentham stepper motor driver, Bentham Instruments, U.K.) have a more basic interface which is capable of receiving commands but not transmitting them. This actually puts more demands on the programmer, because the time required for the instrument to implement coded instructions needs to be taken into account. This is a general feature of writing code for controlling instrumentation: The rate at which the PC works and the rate at which things happen in the rest of the system are different. This is one thing that event-driven programming is very good at, because code can be made to wait for its execution, until a required event.

5.4.7 OSCILLOSCOPE

The oscilloscope is very important to the correct operation of the instrument, although the time-of-flight measurement aspect of the system can operate without it. The oscilloscope is necessary to ensure that the correct pulse waveform is received and that the trigger arrangements for the timer–counter are correct. As indicated in §2.1.2.8, changes in the pulse amplitude and shape can result in timing errors of varying severity. In ultrasound, every disadvantage has the potential to be turned to advantage. In this case, the changing pulse amplitude and shape will often be a result of scattering, which is dependent on particle size and size distribution (Chapter 4). The LeCroy 9450 with waveform processing option is a powerful computer in its own right, with the ability to perform waveform analysis, such as fast Fourier transforms, at high speed.

5.4.7.1 Fourier Analysis

Fourier analysis depends on the property of any time-varying *physical* variable (ξ) that it can be described mathematically as the sum of sine, cosine, and exponential functions of frequency

$$\xi(t) = \sum_{n=1}^{\infty} \xi_n(\omega) \exp(i\omega_n t). \tag{5.7}$$

The coefficients $\xi_n(\omega)$ are called the Fourier coefficients and a plot of these coefficients against frequency is called the *fourier transform* of ξ. In fact, this can be done for any pair of physically related quantities such as (ω, t) and (r, k). The inverse transform $\xi(\omega) \Rightarrow \xi(t)$ is effected in the same way:

$$\xi(\omega) = \sum_{n=1}^{\infty} \xi_n(t) \exp(-i\omega_n t). \tag{5.8}$$

Examples appear in Figures 5.11 and 5.12, for data captured and analyzed with the LeCroy and then transferred to the computer and onto an Excel spreadsheet. The LeCroy manual, to which the reader is referred, contains a great deal of valuable information about fourier transforms and waveform analysis in general.

One general point needs to be made about the choice of digital oscilloscopes for the analysis of pulsed signals such as are used in ultrasound. Because the Fourier components of a pulse rise to much higher frequencies than the center frequency, the bandwidth of the instrument should be very considerably greater than twice the center frequency. The bandwidth should be four times the highest frequency in the pulse. In the case of the 2.25 MHz nominal center frequency transducer which was used to obtain Figure 5.11, a bandwidth of at least 24 MHz is required to represent the pulse accurately.

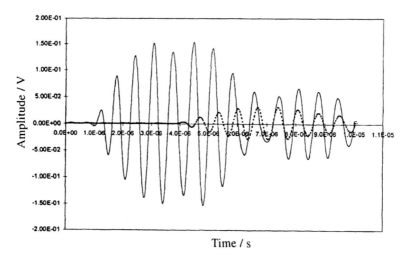

FIGURE 5.11 Pulse waveforms captured at two heights within a creaming emulsion. ———, 6 mm height; – – –, 22 mm height. The solid line corresponds to sound traveling through the serum at the bottom of the tube; the dashed line corresponds to a pulse which has been attenuated in the emulsion. In addition, the pulse is delayed because the velocity is lower in the emulsion than it is in the serum. 2.25 MHz nominal, medium damped transducer with spike excitation.

In Figure 5.11 is shown the waveform captured from the ultrasound profiler at two heights in the emulsion, one at 6-mm height in the serum and the second in the emulsion phase. It is clear that the signal in the serum is much larger than that in the emulsion, and that the signal in the emulsion is also delayed, corresponding to a lower velocity of sound. The nominal center frequency of the pulse is 2.25 MHz, and the fact that there are more than two cycles in the pulse indicates medium damping. Highly damped transducers, which are used for time-of-flight measurements in metals testing equipment, will have only one or at most two cycles in the pulse, to minimize triggering errors in timing circuits. Low damped or undamped transducers would show more than 30 or 40 cycles under these circumstances. Although undamped transducers give a higher output, there is the danger that the resulting long ringing time of the pulse can exacerbate errors resulting from reverberation in the system (§2.1.2.4). Depending on the acoustic path length, the trailing edge of such long pulses can overlap with the reverberations and echoes of the leading edge, creating interference similar to the fringes seen with light scattered from a thin film of oil resting on water. This interference can cause large phase shifts and changes in amplitude and is to be avoided. In the spirit of turning a problem into a solution, a technique called pulse-echo overlap exploits these fringes in order to combine the accuracy of the standing-wave systems (§5.1.5.5) with the speed of pulse-echo systems (§5.1.5.4) (Aurenty *et al.*, 1995; Kaatze *et al.*, 1993).

The amplitude ξ is often defined in terms of a ratio with respect to some arbitrary amplitude ξ_{ref},

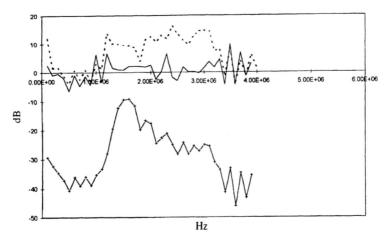

FIGURE 5.12 The Fourier transform of the waveforms in Figure 5.11, plotted as the difference between 10 times the logarithm (base 10) of the water signal and that of the corresponding signal in the emulsion. ———, 6 mm height; – – –, 22 mm height. The lower plot [- + - + -, 10 $\log_{10}$(amplitude in volts) water at all heights in the profiler] is that of the signal in water, to which the other two plots are referred. In the upper two plots, 0 dB represents equality with the amplitude in pure water, a positive value indicates attenuation and a *negative* value indicates a value *higher* than that in water.

$$10 \log_{10}\left(\frac{\xi}{\xi_{ref}}\right), \tag{5.9}$$

whose units are called decibels (dB). The advantage of this is that ratios can be expressed as differences, and decade changes in amplitude appear as 10 dB. Confusingly, the same units are used to express changes in power:

$$10 \log_{10}\left(\frac{\xi^2}{\xi_{ref}^2}\right) = 20 \log_{10}\left(\frac{\xi}{\xi_{ref}}\right) \tag{5.10}$$

The Fourier transform of the signals in Figure 5.11 is shown in Figure 5.12. The lower plot is for pure water and indicates the performance of the transducer. The transducer was excited with a 300-V, 30-ns pulse. If the 6-dB points are taken as a measure of the transducer bandwidth, then the transducer has a frequency of (1.5 ± 0.4) MHz. The nominal frequency of this transducer was 2.25 MHz. Unfortunately, this is a not uncommon experience; commercially available transducers are often poorly characterized and quality comes at a high price. Examination of the transform indicates that the output of the transducer spans a much greater range of frequencies and the 20-dB points are over 2 MHz apart. This bandwidth can be exploited with careful use of the waveform processing capabilities of the oscilloscope. The signal which has passed through the serum (6 mm) is very similar in overall amplitude to that in water, except that it has undergone some interesting phase shifts and the detailed changes in this signal are not understood. The signal through the emulsion has undergone considerable frequency-dependent attenuation, as is evident from a comparison of the shape of the Fourier transform to that of water. In particular, the higher frequencies are much less attenuated than the lower frequencies. No attempt has been made here to interpret these results in terms of scattering theory, primarily because the bandwidth is rather small.

5.4.8 TIMER–COUNTER

The velocity can be determined from the Fourier transform of the phase of the signal, using the oscilloscope. But each waveform record occupies 32 Kbytes of store and it is a relatively slow process, expensive in terms of disk storage, to proceed in this manner. The timer-counter has the advantage of giving a rapid measurement of time-of-flight, which can be corroborated by reference to the oscilloscope from time to time.

Great care needs to be taken in setting the triggering conditions for the timer-counter (§2.1.2.8), and the necessary conditions should first be determined by oscilloscope. The trigger settings should then be applied by software and the resulting time-of-flight checked against the oscilloscope. Since the accuracy of the velocity measurement is dependent on timing, this cross corroboration is important. The instruments involved need to be calibrated according to their manufacturers' instructions. If the transducer is changed, or the method of stimulation altered, the trigger settings need to be rechecked. Transducers immersed in water baths can suffer progressive breakdown (§2.1.2.2) and should be checked regularly. Software control of the instrument ensures a standard and repeatable set of operating conditions, which is difficult to achieve if instruments are set up manually.

5.4.9 THE UVM

Successful ultrasound profiling depends on a knowledge of the temperature dependence of the velocity of sound in the component phases of a dispersion. The ultrasound velocity meter (Cygnus Instruments, Dorchester, U.K.) is a solution to the problem (§2.1), which can be integrated into the ultrasound profiler. This instrument uses a serial interface to communicate with the computer and measures both the velocity of sound and the temperature of the sample simultaneously. Because of the much smaller amount of sample involved (200 ml), temperature dependency can be measured more quickly (a few hours), more accurately (± 1 m s^{-1}, ± 0.1 m s^{-1} precision; $\pm 0.1°C$, $\pm 0.01°C$ precision) and over a wider temperature range ($-20°C$ to $90°C$) than in the profiler. This instrument can be used simultaneously with the profiler if required.

5.4.10 TRANSDUCER EXCITATION

The two common methods of exciting ultrasound transducers are *voltage spike* and *burst rf*. There is not a great deal to choose between the two methods in terms of the overall transfer into acoustic energy. The spike method can be implemented cheaply and produces a broad-band output, causing the transducer to vibrate in all possible modes and frequencies. It can be likened to hitting a very stiff drum with a hammer. It is a cheap and rapid way of exciting a wide range of frequencies and should be used in the case of highly damped transducers where a short pulse is required. The burst rf technique gives much greater control over transducer excitation, since every aspect of the output of equipment such as the Hewlett-Packard 4115A signal generator can be computer controlled. In addition, the relatively large output of 16 V combined with the ability to select the rf frequency to match the resonant frequency of the transducer means that a much more efficient conversion of electrical to acoustic energy can be achieved. This is the reason that apparatus with an amplitude of 16 V can match the acoustic output of a spike generator of 1 kV. However, if a range of frequencies are required, then the equipment must be set to each frequency separately, a much slower process than can be achieved with spike generation.

5.4.11 CABLING

A common source of error, which can be difficult to track down, is faulty cabling and connections. This should be the first place to check if problems are suspected. BNC connectors are particularly prone to break down; a visual inspection will not detect the most common faults in these connectors and cabling. Continuity testing of the conductor and shield should be carried out separately to confirm correct operation.

5.4.12 CALIBRATION

The overall accuracy depends on careful and regular calibration with double-distilled water. Calibration can be automated and any drift in the apparatus, its transducers, and associated electronics can be checked against the initial calibration of the apparatus. This is the most effective way to check the performance of the system.

5.4.13 SAMPLE CHANGER

Emulsions can take a great deal of time to cream, so the ability to change samples during an experiment is valuable. This is particularly so because creaming can be affected by the thermal and elastic stresses arising from sample movement. The entire apparatus is held in a water bath to ensure temperature uniformity, and temperature equilibration can take some time. Often a complex sequence of measurements are required, and software control of the sample changer permits this. The sample changer comprises a carousel of six sample tubes, driven by a stepper motor. The computer controls the movement of the probes up and down each sample tube through a second stepper motor and synchronizes probe movement with the carousel movement. In this way, experiments can be set up which can run for hours, days, weeks, or even months.

5.4.14 TEMPERATURE CONTROL

A high degree of temperature uniformity and accuracy is required for successful ultrasound profiling. Since temperature changes are generally avoided during experiments, temperature is controlled by an independent thermostat, which is set manually. Temperature logging can easily be incorporated, if necessary, by including a temperature measurement device on the IEEE bus.

5.4.15 DATA STORAGE AND ANALYSIS

Automatic control of ultrasound experiments can generate large amounts of data very quickly. Since the intention of programming is to make ultrasound instrumentation accessible to scientists in general, many different people may be using the instrument. Careful control of data storage is important in this case.

A spreadsheet application such as Excel is used to log the data from an experiment and to display a chart of the results during the experiment. This is a very convenient and successful approach. Data are transferred to the spreadsheet using a method called dynamic data exchange (DDE), which is driven by the VB control program. The spreadsheet is driven by macro commands from the VB control program. Thus, the user does not have to know how to use Excel. A template is produced; the VB control program opens it and transfers the data to it. A chart, incorporated into the template, automatically displays the results as they are transferred to the spreadsheet. The template can be edited if required. Thus, the user can modify the programming and introduce additional algorithms in the form of macros or VB code which is incorporated into the spreadsheet. The template incorporates links to a set of spreadsheets created during a separate calibration procedure. The data from the calibrations are automatically linked to the current data so that the velocity is automatically calculated. This procedure permits a separate calibration for each cell, at 1-mm steps up each tube. The sheet is automatically named and saved by the VB control program, allowing the automatic organization of the storage of the data. If the computer is part of a network such as Windows NT, the spreadsheet can be stored on a file server and automatically backed up to tape. The data can be secured by such a system, as can access to the ultrasound instrument. Networking the data allows it to be viewed over the network, permitting the progress of an experiment to be viewed from anywhere in the world. In this way, data from experiments can be linked directly into documents, such as this book.

5.4.16 CONCLUSION

The future success of ultrasound instrumentation is very much dependent on its successful integration with computer techniques and electronics. The examples shown here demonstrate that the technology now exists to produce a new generation of ultrasound instruments.

Appendix A

Basic Theory

This outline of the basic theory is meant to guide the reader; it is not a rigorous treatment of the subject of ultrasonic propagation. Before this appendix, Chapter 1 and Chapter 2, §2.1 should be read.

A sound wave causes local compression and extension in the material through which it passes. In this derivation of the velocity of sound it is hypothesized that propagation in one dimension is an accurate analogy for propagation in liquids. Suppose that a material segment of width Δx, area A and bulk modulus B is stretched by a force F to a width of $\Delta z + \Delta x$ (see Figure A.1).

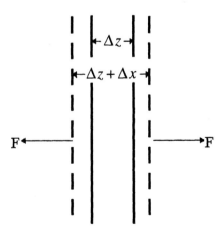

FIGURE A.1 Diagrammatic representation of the expansion phase of the passage of a sound wave.

The strain in this segment is then given by $\Delta x/\Delta z$. The stress is given, from the definition of the bulk modulus, by:

$$\frac{F}{A} = B\frac{\Delta x}{\Delta z}. \quad (A.1)$$

The excess force ΔF that causes an infinitely thin segment Δz of the material to accelerate is given by:

$$\Delta F = A \times B \times \left(\frac{\partial^2 x}{\partial z^2}\right) \times \Delta z. \quad (A.2)$$

This force is given, from Newton's second law, by the product of mass $(\rho A \Delta z)$ and acceleration $(\partial^2 x/\partial t^2)$ so that the acceleration can be written as

$$\frac{\partial^2 x}{\partial t^2} = \frac{B}{\rho} \times \frac{\partial^2 x}{\partial z^2}. \quad (A.3)$$

The velocity of sound v for a liquid is then given by

$$v = \frac{\partial z}{\partial t} = \sqrt{\frac{B}{\rho}} = \sqrt{\frac{1}{\kappa\rho}}. \quad (A.4)$$

This is the Wood equation (Equation 2.4). The derivation for mixtures and heterogeneous liquids follows in Chapter 2, §2.2. A strict derivation of the propagation equations, from scattering theory, is given in Chapter 4. The effects of high attenuation on propagation are considered in Chapter 2, §2.3.

A more general derivation of equations for velocity and attenuation may be obtained from scattering theory, via the generalized dispersion relation Equation 4.22 and the stress/strain-rate relationship Equation 4.15. By combining these equations with Equation 4.16 or 4.19 we obtain the following results for an isotropic elastic solid, a pure Newtonian fluid, a mixture of Newtonian fluids, and a viscoplastic material. *Note that the following results are derived from the dispersion relation for single-particle scattering and cannot be generalized to mixtures without error,* nor can the approach adopted in the following section be regarded as rigorous. Strictly, the single-particle scattering coefficients should be derived for the case in question and the result inserted in the dispersion relation for multiple scattering, Equation 4.41 or one of its simplified versions such as Equation 4.47.

A.1 THE ISOTROPIC ELASTIC SOLID

From Equation 4.16 for the solid case and assuming that $e_k \gg 1$ and $f_k \ll 1$ (this is equivalent to the adiabatic approximation, §4.3.2.6, Equations 4.5 and 4.6), we obtain the following results for an isotropic elastic solid:

$$\frac{1}{k^2} \cong \frac{v^2}{2\omega^2}\{1 - ie_k + (1 - ie_k)\} \cong \frac{v^2}{\omega^2}(-ie_k), \quad (A.5)$$

APPENDIX A

where

$$e_k = \frac{(\mu + \lambda + \tfrac{2}{3}\mu)i\omega}{\rho v^2} = \frac{(B + \tfrac{4}{3}G)i\omega}{\rho v^2} \quad (A.6)$$

so that

$$v = \sqrt{\frac{(\lambda + 2\mu)}{\rho_1}} = \sqrt{\frac{(B + \tfrac{4}{3}G)}{\rho_1}}, \quad (A.7)$$

which is the standard result for an isotropic elastic solid.

A.2 THE PURE NEWTONIAN FLUID

Using Equation 4.19 for the liquid case and assuming that $e_k \ll 1$ and $f_k \ll 1$, we have

$$\frac{1}{k^2} \cong \frac{v^2}{\omega^2}\{1 - ie_k - (\gamma - 1)f_k\}. \quad (A.8)$$

Using the results for e_k and f_k from Equation 4.23, we obtain

$$k' = \frac{\omega}{v} \quad \text{where} \quad v = \sqrt{\frac{B}{\rho_1}} = \sqrt{\frac{1}{\kappa_{a1}\rho_1}} \quad (A.9)$$

for a *pure* liquid which is the usual equation for the velocity of sound, and

$$k'' = \alpha = \frac{\omega}{2v^3}\left\{\frac{(\eta_v + \tfrac{4}{3}\eta_s)}{\rho_1} + \sigma(\gamma - 1)\right\} \quad (A.10)$$

for the attenuation.

A.3 MIXTURE OF TWO PURE NEWTONIAN FLUIDS

Equations A.8 and A.9 do not apply in the case of mixtures of different liquids, and in these cases v is given by Equation 2.34 §2.4.5 (the modified Urick equation), for solutions.

A.4 VISCOPLASTIC MATERIALS

Here we assume that $f_k \ll 1$. This gives the following simplified dispersion relation:

$$\frac{1}{k^2} = \frac{v^2}{\omega^2}(1 - ie_k). \quad (A.11)$$

We now assume that the Lamé constants are both complex and frequency-dependent:

$$e_k = \frac{\left(\tilde{B}(\omega) + \tfrac{4}{3}G(\omega)\right)i\omega}{\rho v^2}, \tag{A.12}$$

where

$$B(\omega) = \frac{1}{\kappa_{al}} + \tilde{B}(\omega) = B' + B'' \text{ and } G(\omega) = G' + G'' = \mu(\omega) \tag{A.13}$$

and

$$\tilde{B}(\omega) = \lambda(\omega) + \frac{2}{3}\mu(\omega) \tag{A.14}$$

We now *define* a complex elastic modulus $M(\omega)$ such that

$$M(\omega) = \lambda(\omega) + \frac{5}{3}\mu(\omega) = B(\omega) + \frac{4}{3}G(\omega) = M' + iM'' \tag{A.15}$$

so that we can write the complex velocity of sound as

$$v(\omega) = \sqrt{\frac{M(\omega)}{\rho_i}}. \tag{A.16}$$

Then the phase velocity will be

$$v = \sqrt{\frac{2\sqrt{M'^2 + M''^2}}{\rho_1\left(1 + \dfrac{M'}{\sqrt{M'^2 + M''^2}}\right)}} \tag{A.17}$$

and the attenuation,

$$\alpha = \omega\sqrt{\rho_1 \frac{1 - \dfrac{M'}{\sqrt{M'^2 + M''^2}}}{\sqrt{M'^2 + M''^2}}}. \tag{A.18}$$

These equations may be compared with the results quoted without proof in Pethrick (1983, p. 265).

If the attenuation is small enough that $k' \gg k''$, then

$$v = \sqrt{\frac{M'}{\rho_1}} = \sqrt{\frac{B' + \tfrac{4}{3}G'}{\rho_1}} \cong \sqrt{\frac{B'}{\rho_1}} \tag{A.19}$$

$$\alpha = \frac{\omega M''}{2M'}\sqrt{\frac{\rho_1}{M'}} = \frac{B'' \tfrac{4}{3}G''}{2\rho_1 v^3}\omega. \tag{A.20}$$

Appendix B

MathCad Solutions of the Explicit Scattering Expressions

Allegra and Hawley (1972) give explicit expressions for the scattering which are valid in the long wavelength limit $\lambda \gg d$ (see also Chapter 4). These expressions permit the use of a package such as MathCad to analyze the scattering. In this section is written a solution of the scattering problem which both presents the explicit solutions as detailed by Allegra and Hawley and lays the solution out in a way that can be copied directly into a MathCad sheet. The solution is presented for a hexadecane-in-water emulsion, and the data employed appears in the Table B.1. The MathCad sheet on which this is based was originally written by Julian McClements and has been extensively revised by the author and by Valerie Pinfield, who also collected the data for the thermoelastic and viscoinertial terms. Where results are referred to in the book, the figures have been edited and re-presented in the body of the book. It is not a difficult matter to reproduce these calculations, which are very useful in analyzing the contribution of scattering to ultrasound propagation.

Although the data presented applies to hexadecane-in-water, readers may apply the same sheet to any dispersed system comprising liquid particles by obtaining the appropriate data. The subscripting and symbols have been transcribed directly from the MathCad sheet and may not correspond exactly to the symbols used in the rest of the book. This sheet is intended to be read in conjunction with Chapter 4, where the variables used are defined.

Define the imaginary number $j = (-1)^{0.5}$.

Calculate the volume average values of density and compressibility:

$$\rho_0 := \rho_1(1 - \phi) + \rho_2\phi \qquad \rho_0 = 953.1$$

$$\kappa_0 := \kappa_1(1 - \phi) + \kappa_2\phi \qquad \kappa_0 = 5.052 \times 10^{-10}$$

TABLE B.1 Thermophysical Properties of Component Phases

Property at 20°C	Water	Hexadecane
Velocity (m s^{-1})	$v_1 := 1482$	$v_2 := 1358$
Density (kg m^{-3})	$\rho_1 := 998$	$\rho_2 := 773.5$
Attenuation (Neper m^{-1})	$\alpha_1 := 0.4$	$\alpha_2 := 1.7$
Viscosity (Pa s)	$\mu_1 := 0.001$	$\mu_2 := 0.00344$
Specific heat (J kg^{-1} m^{-3})	$C_1 := 4182$	$C_2 := 2092$
Thermal conductivity (W m^{-1} K^{-1})	$\tau_1 := 0.591$	$\tau_2 := 0.14$
Coefficient of cubical expansion	$\beta_1 := 0.00021$	$\beta_2 := 0.00091$
Compressibility	$\kappa_1 := \dfrac{1}{v_1 \cdot v_1 \cdot \rho_1}$	$\kappa_2 := \dfrac{1}{v_2 \cdot v_2 \cdot \rho_2}$
Volume fraction	$\phi := 0.2$	$(1 - \phi) = 0.8$
Radius		$r := 1.10^{-6}$

Calculate the parameters to be used in the equations:

$$\log f := 2.20 \qquad f_{\log f} := 10^{0.5 \log f} \qquad \omega_{\log f} := 2 \times 3.1415 \times f_{\log f}$$

$$\delta s_{\log f} := \left(2 \times \frac{\mu 1}{\omega_{\log f} \times \rho 1}\right)^{0.5}$$

$$\delta t1_{\log f} := \left(2 \times \frac{\tau 1}{\omega_{\log f} \times \rho 1 \times C1}\right)^{0.5} \qquad \delta t2_{\log f} := \left(2 \times \frac{\tau 2}{\omega_{\log f} \times \rho 2 \times C2}\right)^{0.5}$$

$$S_{\log f} := 9 \times \frac{\delta s_{\log f}}{4r}\left(1 + \frac{\delta s_{\log f}}{r}\right) \qquad T_{\log f} := \frac{1}{2} + 9 \times \frac{\delta s_{\log f}}{4r}$$

$$b1_{\log f} := \frac{r}{\delta t1_{\log f}}(1 + j) \qquad b2_{\log f} := \frac{r}{\delta t2_{\log f}}(1 + j)$$

$$Z := (1 - jb1_{\log f}) \frac{\sin b2_{\log f}}{\sin b2_{\log f} - b2_{\log f} \cos b2_{\log f}} \left(= \frac{1 - jb1_{\log f}}{1 - b2_{\log f} \cot b2_{\log f}} \right) \cdot$$

Note: Exploring the limit

$$\text{As } r\omega^{1/2} \to 0, \quad b_1, b_2 \to 0 \ ,$$

we obtain the following:

$$\cot b_2 = \frac{1}{b_2} - \frac{b_2}{3} - \frac{b_2^3}{45} + \cdots$$

APPENDIX B

$$\therefore \lim_{b_2 \to 0}\{\cot b_2\} \to \frac{1}{b_2} - \frac{b_2}{3}$$

$$\lim_{r\omega^{1/2} \to 0}\{Z\} = \frac{1-jb_1}{1-b_2\left[\frac{1}{b_2} - \frac{b_2}{3}\right]} = \frac{1-jb_1}{1-1+\frac{b_2^2}{3}} = 3\frac{1-jb_1}{b_2^2}.$$

This closely agrees with the MathCad calculation of the limit using the full explicit expressions.

Continuing with the analysis:

$$F_{\log f} := \frac{1 - \frac{\tau 1}{\tau 2} \times Z_{\log f}}{1 - j \times b1_{\log f}} \quad (B.1)$$

Note: Checking the limiting analytical expressions with the MathCad calculation:

$$\lim_{r\omega^{1/2} \to 0}\{F\} = \frac{1 - \frac{\tau_1}{\tau_2}\lim_{r\omega^{1/2} \to 0} Z}{1-jb_1} = \frac{1 - \frac{\tau_1}{\tau_2} 3\frac{1-jb_1}{b_2^2}}{1-jb_1}. \quad (B.2)$$

Using Equation B.1:
Note: Checking the limiting analytical expressions with the MathCad calculation: Equation B.2 then gives the following result:

$$\lim_{r\omega^{1/2} \to 0}\left\{\frac{1}{b_1^2 F}\right\} = \frac{1-jb_1}{b_1^2\left[1 - \frac{\tau_1}{\tau_2} 3\frac{1-jb_1}{b_2^2}\right]} = \frac{-\rho_2 C_2}{3\rho_1 C_1}.$$

This closely agrees with the MathCad calculation of this limit. Note in the following excerpt from the MathCad sheet that the imaginary part of the expression $1/b_1^2 F$ gives the form of the attenuation and its real part gives the form of the velocity.

Calculating the effective density and adiabatic compressibility:

$$\rho_{\log f} := \rho_0 - \Re\left(\left\{\frac{(\rho 2 - \rho 1)^2}{\rho 2 + \rho 1 \times T_{\log f} + j \times \rho 1 \times s_{\log f}}\right\} \times \phi\right)$$

$$\kappa_{\log f} := \kappa 0 - \Re\left(3 \times \kappa 1 \times \phi \times \frac{y-1}{(b1_{\log f})^2 \times F_{\log f}}\left(1 - \beta 2 \times \rho 1 \times \frac{Ca}{\beta 1 \times \rho 2 \times C2}\right)^2\right).$$

Calculating the velocity and attenuation:

$$c_{\log f} := \left(\rho_{\log f} \kappa_{\log f}\right)^{-0.5}$$

$$\alpha_{\log f} := \frac{c1}{f_{\log f}} \Im m\left(\omega_{\log f} \left(\rho_{\log f} \kappa_{\log f}\right)^{0.5}\right)$$

The results of these calculations may be seen in Figures 4.9, 4.10, and 4.11.

Glossary

1	continuous phase subscript
2	dispersed phase subscript
A	vector potential
A_0	zero-order scattering coefficient for the acoustic propagational mode
A_1	first-order scattering coefficient of the propagational acoustic mode
A_ψ	azimuthal component of **A**
A_n, B_n, C_n	single-particle scattering coefficients in the continuous phase
A'_n, B'_n, C'_n	single-particle scattering coefficients in the interior of the particle
$Av, Bv, Cv, Dv, Ev, Fv, A\rho, B\rho, C\rho, D\rho, E\rho, F\rho$	empirical coefficients relating velocity and density to concentration and temperature in sugar solutions
a	total number of carbon atoms in the triacylglycerol molecule
a	parameter in the renormalization method
a_n	modified scattering coefficients for the propagational acoustic mode
B	adiabatic bulk modulus ($= 1/\kappa_a$)
$\tilde{B}(t)$	time-dependent bulk modulus (omits the adiabatic bulk modulus)
b	total number of unsaturated bonds in the triacylglycerol molecule

w	mass fraction
X_z	reactive component of impedance Z
x	sugar concentration in g 100ml^{-1}
x	mole fraction
x	distance
Z	impedance (electrical or acoustical or electromechanical)
α	attenuation coefficient
α	Urick equation first order parameter
α_T	preexponential temperature coefficient of velocity in oils and fats not undergoing crystallization
α_s	excess attenuation (due to the addition of scatterers), scattering contribution to the total attenuation
α_e	Urick equation first-order parameter as determined by experiment (the first-order effective coefficient)
α_1	attenuation measured in the continuous phase (in the absence of scatterers)
α_{total}	total attenuation as measured
β	volume coefficient of thermal expansivity
β_T	exponential temperature coefficient of velocity in oils and fats not undergoing crystallization
γ	ratio of the specific heats
Δ	dilatation
$\Delta(1/v^2)$	parameter in the renormalization method
ΔH_f	enthalpy change per mole of crystallizing material
Δp	pressure deviation
Δp_0	maximum pressure deviation
Δv_{exp}	experimental error in velocity
$\Delta \phi_{hyd}$	volume change due to hydration in a protein
δ	Urick equation second-order parameter
$\delta(t)$	Dirac delta function
δ_e	Urick equation second-order parameter as determined by experiment (second-order effective coefficient)
δ_{ij}	Kronecker delta symbol
δ_s	shear decay length
δ_t	thermal skin depth, thermal decay length
ϵ	permittivity
ζ	phase of a wave
ζ	zeta potential
ζ_{ij}	cross scattering coefficient in the attenuation
η_s	shear viscosity
η_v	bulk viscosity
κ	compressibility
$\overline{\kappa_2}$	partial specific adiabatic compressibility of solute
κ_a	adiabatic compressibility of pure material or of mixture

GLOSSARY

κ_{a1}	adiabatic compressibility of the continuous phase ($=1/B$)
κ_{a2}	adiabatic compressibility of the dispersed phase
λ	wavelength
λ	first Lamé constant
μ	second Lamé constant
μ_d	dynamic mobility
θ	azimuthal angle in spherical coordinate system
θ	parameter in the modified Urick equation
ξ	displacement of a volume element of material
ρ	density, instantaneous density
ρ_0	static density, original density
σ	thermometric conductivity, also known as thermal diffusivity
$\sigma_{\bar{r}}$	standard deviation of log-normal particle-size distribution
τ	thermal conductivity
τ_r	relaxation time
Φ_s	volume fraction of crystallizable material in a dispersion, volume fraction of dispersed phase
Φ	scattering coefficient parameter
ϕ	angle in spherical coordinate system
ϕ	volume fraction
ϕ_c	sum of the constitutive or group volumes (proteins)
$\phi_i(r_i)$	volume fraction occupied by the ith fraction of the particle-size-distribution
ϕ_{cav}	volume of the cavity in a protein
ϕ_{scatt}	volume fraction as determined by the scattering equation (Equation 4.56)
ϕ_{Urick}	volume fraction as determined by the Urick equation (Equation 4.54)
ϕ_{V_2}	"apparent" specific volume of solute
ϕ_{Vm}, ϕ_{VM}	"apparent" specific molar volume or molar volume
ϕ_{κ_2}	"apparent" compressibility
$\phi_{\kappa M}$	"apparent" molar adiabatic compressibility
χ_i	first coefficient of the attenuation–concentration relationship
φ_0	wave potential
φ_R	potential of the scattered compressional acoustic wave
φ_t	potential of the scattered thermal wave
ν	kinematic viscosity ($=\eta_s/\rho$)
ω	radial frequency ($=2\pi f$)
ω_M	Minnaert frequency
ω_r	relaxation frequency
ς	second coefficient of the attenuation–concentration relationship

BIBLIOGRAPHY

A.O.C.S. (1973) Official Method Cd 10-57.

Akulichev, V. A., and Bulanov, V. N. (1982). Sound Propagation in a Crystallizing Liquid. *Soviet Physics—Acoustics (English Translation)* **27**, 377–381.

Alba, E. (1992). Method and Apparatus for Determining Particle Size Distribution and Concentration in a Suspension Using Ultrasonics. *U.S. Patent No. 5121629.*

Allegra, J. R., and Hawley, S. A.(1972). Attenuation of Sound in Suspensions and Emulsions: Theory and Experiments. *Journal of the Acoustical Society of America* **51**, 1545–1564.

Almond, D. P., and Patel, P. M. (1996). *Photothermal Science and Techniques.* Chapman and Hall, London.

Anson, L. W., and Chivers, R. C. (1993). Ultrasonic Scattering from Spherical-Shells Including Viscous and Thermal Effects. *Journal of the Acoustical Society of America* **93**, 1687–1699.

Antosiewicz, J., and Shugar, D. (1984). Hydration of Alcohols by Ultrasonic Velocity Measurements in Ternary Systems. *Journal of Solution Chemistry* **13**, 493–503.

Apfel, R. E. (1981). Acoustic Cavitation. In *Methods of Experimental Physics: Ultrasonics* (Series Editors: Marton, L., and Marton, C.; Volume Editor: Edmonds, P.D.), Academic Press, San Diego, Vol. 19, pp. 355–411.

Archer, G. P., Kennedy, C. J., and Povey, M. J. W. (1996). Investigations of Ice Nucleation in Water-in-Oil Emulsions Using Ultrasound Velocity Measurements. *Cryoletters* **17**, 391–396.

Atkins, P. W. (1982). *Physical Chemistry,* 2nd Edition. Oxford University Press, Oxford, U.K.

Attwood, D., Johansen, L., Tolley, J. A., and Rassing, J. (1981). A New Ultrasonic Method for the Measurement of the Diffusion Coefficient of Drugs within Hydrogel Matrices. *International Journal of Pharmaceutics* **9**, 285–294.

Audebrand, M., Doublier, J. L., Durand, D., and Emery, J. R. (1995). Investigation of Gelation Phenomena of Some Polysaccharides by Ultrasonic Spectroscopy. *Food Hydrocolloids* **9**, 195–203.

Aurenty, P., Schröder, A., and Gandini, A. (1995). Water-in-Alkyd-Resin Emulsions: Droplet Size and Interfacial Tension. *Langmuir* **11**, 4712–4718.

Bachman, S., Klimaczak, B., and Gasnya, Z. (1978). Non-Destructive Viscometric Studies of Enzymic Milk Coagulation. *Acta Alimentaria Polonica* **4**, 55–62.

Bae, J. R. (1996). Ultrasonic Study of Gelation Process in Egg-white Proteins. *Japanese Journal of Applied Physics. Part 1. Regular Papers and Short Notes* **35**, 2934–2938.

Ballaró, S., Mallamace, F., and Wanderlingh, F. (1980). Sound Velocity and Absorption in Microemulsion. *Physics Letters* **A77**, 198–202.

Barrett-Gültepe, M. A., Gültepe, M. E., and Yeager, E. B. (1983). Compressibility of Colloids. 1. Compressibility Studies of Aqueous Solutions of Amphiphilic Polymers and Their Adsorbed State on Polystyrene Latex Dispersion by Ultrasonic Velocity Measurements. *Journal of Physical Chemistry* **87**, 1039–1045.

Barrett-Gültepe, M. A., Gültepe, M. E., McCarthy, J. L., and Yeager, E. B. (1988). The Determination of Particle-Size Distribution of Perfluorochemical Emulsions by Ultrasonic Measurements. *Biomaterials Artificial Cells and Artificial Organs* **16**, 691–692.

Barrett-Gültepe, M. A., Gültepe, M. E., McCarthy, M. J., and Yeager, E. B. (1989). A Study of Steric Stability of Coal Water Dispersions by Ultrasonic-Absorption and Velocity-Measurements. *Journal of Colloid and Interface Science* **132**, 144–160.

Batchelor, G. K. (1967). *An Introduction to Fluid Dynamics*. Cambridge University Press, U.K.

Benguigui, L., Emery, J. R., Durand, D., and Busnel, J. P. (1994). Ultrasonic Study of Milk-Clotting. *Lait* **74**, 197–206.

Berchiesi, G., Amico, A., Vitali, G., Amici, L., and Litargini, P. (1987). Ultrasonic Investigation in Aqueous Solutions of Sucrose. *Journal of Molecular Liquids* **33**, 157–181.

Beyer, R. T., and Letcher, S. V. (1969). *Physical Ultrasonics*. Academic Press, San Diego.

Bhattacharya, A. C., and Deo, B. B. (1981). Ultrasonic Propagation in Coconut Oil in the Vicinity of Phase Transition. *Indian Journal of Pure and Applied Physics* **19**, 1172–1177.

Biot, M. A. (1956a). Theory of Propagation of Elastic Waves in a Fluid-Saturated Porous Solid. I. Low-Frequency Range. *Journal of the Acoustical Society of America* **28**, 168–178.

Biot, M. A. (1956b). Theory of Propagation of Elastic Waves in a Fluid-Saturated Porous Solid. II. Higher-Frequency Range. *Journal of the Acoustical Society of America* **28**, 179–191.

Bonnet, J. C., and Tavlarides, L. L. (1987). Ultrasonic Technique for Dispersed-Phase Holdup Measurements. *Industrial and Engineering Chemistry Research* **26**, 811–815.

Borthakur, A., and Zana, R. (1987). Ultrasonic Absorption Studies of Aqueous Solutions of Nonionic Surfactants in Relation with Critical Phenomena and Micellar Dynamics, *Journal of Physical Chemistry*, **91**, 5957–5960.

Bouts, M. N., Wiersma, R. J., Muijs, H. M., and Samuel, A. J. (1995). An Evaluation of New Asphaltene Inhibitors: Laboratory Study and Field Testing. *Journal of Petroleum Technology* **September**, 782–787.

Breazeale, M. A., Cantrell, J. H., and Heyman, J. S. (1981). Ultrasonic Wave Velocity and Attenuation Measurements. In *Methods of Experimental Physics: Ultrasonics* (Series Editors: Marton, L., and Marton, C.; Volume Editor: Edmonds, P.D.), Academic Press, San Diego, Vol. 19, pp. 67–137.

Buckin, V., Kankiya, B., and Kazaryan, R. (1989). Hydration of Nucleosides in Dilute Aqueous-Solutions Ultrasonic: Velocity and Density-Measurements. *Biophysical Chemistry* **34**, 211–223.

Bulanov, V. A. (1979). Crystallization of a Supercooled Liquid in a Sound Field: Steady-State Dynamics of Crystalline-Phase Nuclei in the Liquid. *Soviet Physics—Acoustics (English Translation)* **25**, 202–207.

Buttner, B., Owusu-Apenten, R., and Povey, M. J. W. (1996). ANS-Fluorescence and Ultrasonic Velocimetry Analysis of the Heat Effect on Ovalbumin Solutions. Internal Report, University of Leeds, U.K.

Cao, Y., Dickinson, E., and Wedlock, D. J. (1988). Influence of Polysaccharides on the Creaming of Casein-Stabilized Emulsions. *Food Hydrocolloids* **5**, 443–454.

Cao, Y., Dickinson, E., and Wedlock, D. J. (1990). Creaming and Flocculation in Emulsions Containing Polysaccharide. *Food Hydrocolloids* **4**, 185–195.

Cao, Y., Dickinson, E., and Wedlock, D. J. (1991). Influence of Polysaccharides on the Creaming of Casein-Stabilized Emulsions. *Food Hydrocolloids* **5**, 443–454.

Carter, C., Hibberd, D. J., Howe, A. M., Mackie, A.R., and Robins, M. M. (1986). Non-intrusive Determination of Particle Size Distribution in a Concentrated Dispersion. *Progress in Colloid and Polymer Science* **76**, 37–41.

Chabance, B., Perrotin, B., Fiat, A. M., Miglioresamour, D., Jolles, P., Guillet, R., and Boynard, M. (1996). Measurement of Human Platelet Microaggregates by a New Method—Ultrasonic Interferometry. *Journal of Laboratory and Clinical Medicine* **127**, 296–302.

Chalikian, T. V., Gindikin, V. S., and Breslauer, K. J. (1995). Volumetric Characterizations of the Native, Molten Globule and Unfolded States of Cytochrome-C at Acidic pH. *Journal of Molecular Biology* **250**, 291–306.

Charlier, J. P., and Crowet, F. (1986). *Journal of the Acoustical Society of America* **79**, 895–900.

Chivers, R. C., Aebischer, H. A., Horne, S. A., and Ennos, A. E. (1992) Vibrations of an Annular Membrane. In *Developments in Acoustics and Ultrasonics* (Editors: Povey, M. J. W., and McClements, D. J.), Institute of Physics, Bristol, U.K., pp. 225–232.

Choi, P. K., Bae, J. R., and Takagi, K. (1987). Gelation and Ultrasonic Hysteresis in Egg-White. *Japanese Journal of Applied Physics. Part 1. Regular Papers and Short Notes* **26**, 32–34.

Chu, B., Zhou, Z., Wu, G., and Farrell, H. M. (1995). Laser Light Scattering of Model Casein Solutions: Effects of High Temperature. *Journal of Colloid and Interface Science* **170**, 102–112.

Colladon, J. D., and Sturm, J. C. F. (1827). Memoir on the Compression of Liquids. *Annales de Chimie (Paris)* **36**, 225–227.

Commander, K. W., and Prosperetti, A. (1989). Linear Pressure Waves in Bubbly Liquids: Comparison Between Theory and Experiments. *Journal of the Acoustical Society of America* **85**, 732–746.

Contreras Montes de Oca, N. I., Fairley, P., McClements, D. J., and Povey, M. J. W. (1992). Analysis of the Sugar Content of Fruit Juices and Drinks Using Ultrasonic Velocity Measurements. *International Journal of Food Science and Technology* **27**, 515–529.

Conway, B. E., and Verral, R. E. (1966). Partial Molar Volumes and Adiabatic Compressibilities of Tetraalkyllammonium and Aminium Salts in Water. I. Compressibility Behaviour. *Journal of Physical Chemistry* **70**, 3952–3961.

Cosgrove, N., O'Donnell, C., and Povey, M. J. W. (1996). Unpublished results, University of Leeds.

Coupland, J., Dickinson, E., McClements, D. J., Povey, M. J. W., and de Rancourt de Mimmerand, C. (1993). Crystallization in Simple Paraffins and Monoacid Saturated Triacylglycerols Dispersed In Water. In *Food Colloids and Polymers: Stability and Mechanical Properties* (Editors: Dickinson, E., and Walstra, P.), Royal Society of Chemistry, Cambridge, U.K., pp. 243–249.

d'Agostino, L., and Brennen, C. E. (1988). Acoustical Absorption and Scattering Cross-Sections of Spherical Bubble Clouds. *Journal of the Acoustical Society of America* **84**, 2126–2134.

D'Angelo, M., Onori, G., and Santucci, A. (1994). Self-Association of Monohydric Alcohols in Water: Compressibility and Infrared Absorption Measurements. *Journal of Chemistry and Physics* **100**, 3107–3112.

D'Arrigo, G., and Paparelli, A. (1987). Sound Propagation in Water-Ethanol Mixtures at Low Temperatures. I. Ultrasonic Velocity. *Journal of Chemical Physics* **88**, 405–415.

de Boer, R. B., Leerlooyer, K., Eigner, M. R. P., and van Bergen, A. R. D. (1995). Screening of Crude Oils for Asphalt Precipitation: Theory, Practice and the Selection of Inhibitors. *SPE Production and Facilities* **February**, 55–61.

DeHoop, A. T. (1995). *Handbook of Radiation and Scattering of Waves.* Academic Press, London.

del Grosso, V. A., and Mader, C. W. (1972). Velocity of Sound in Pure Water. *Journal of the Acoustical Society of America* **52**, 1442–1445.

Dickinson, E., and McClements, D. J. (1995b). Ultrasonic Characterization of Food Colloids. In *Advances in Food Colloids.* Chapman and Hall, London, pp. 176–210.

Dickinson, E., and McClements, D. J. (1995a). *Advances in Food Colloids.* Chapman and Hall, London, pp. 102–144.

Dickinson, E., Goller, M. I., McClements, D. J., Peasgood, S., and Povey, M. J. W. (1990). Ultrasonic Monitoring of Crystallization in an Oil-in-Water Emulsion. *Journal of the Chemical Society, Faraday Transactions* **86**, 1147–1148.

Dickinson, E., McClements, D. J., and Povey, M. J. W. (1991). Ultrasonic Investigation of the Particle Size Dependence of Crystallization in n-Hexadecane-Water Emulsions. *Journal of Colloid and Interface Science* **142**, 103–110.

Dickinson, E., Goller, M. I., and Wedlock, D. J. (1993a). Creaming and Rheology of Emulsions Containing Polysaccharide and Non-Ionic or Anionic Surfactants. *Colloids and Surfaces A: Physiochemical and Engineering Aspects* **75**, 195–201.

Dickinson, E., Kruizenga, F., Povey, M. J. W., and van der Molen, M. (1993b). Crystallization in Oil-in-Water Emulsions Containing Liquid and Solid Droplets. *Colloids and Surfaces A: Physiochemical and Engineering Aspects* **81**, 273–279.

Dickinson, E., Ma, J., and Povey, M. J. W. (1994). Creaming of Concentrated Oil-in-Water Emulsions Containing Xanthan. *Food Hydrocolloids* **8**, 481–497.

Dickinson, E., Ma, J., and Povey, M. J. W. (1996). Crystallization Kinetics in Oil-in-Water Emulsions Containing a Mixture of Solid and Liquid Droplets. *Journal of the Chemical Society, Faraday Transactions 1* **92**, 1213–1215.

Dobromyslov, N. A., and Koshkin, N. I. (1970). Velocity of Sound in Molecular Crystals Near the Melting Point. *Soviet Physics—Acoustics (English Translation)* **15**, 386–387.

Domenico, S. N. (1982). Acoustic Wave Propagation in Air-Bubble Curtains in Water—Part I: History and Theory. *Geophysics* **47**, 345.

Edmonds, P. D., Volume Editor (1981). *Methods of Experimental Physics: Ultrasonics* (Series Editors: Marton, L., and Marton, C.), Academic Press, San Diego, Vol. 19.

Eggers, F. G., and Kaatze, U. (1996). Broad-Band Ultrasonic Measurement Techniques for Liquids. *Measurement Science and Technology* **7**, 1-19.

Eggers, F. G., and Funck, T. (1973). *Review of Scientific Instruments* **44**, 969–977.

Elias, J. G., and Eden, D. (1979). High-Accuracy Differential Measurement of Ultrasonic Velocity in Liquids. *Review of Scientific Instruments* **50**, 1299–1302.

Epstein, P. S., and Carhart, R. R. (1953). The Absorption of Sound in Suspensions and Emulsions. I. Water Fog in Air. *Journal of the Acoustical Society of America* **25**, 553–565.

Esquivelsirvent, R., Tan, B., Abdelraziq, I., Yun, S., and Stumpf, F. (1993). Absorption and Velocity of Ultrasound in Binary-Solutions of Poly(Ethylene Glycol) and Water. *Journal of the Acoustical Society of America* **93**, 819–820.

Evans, J. M., and Attenborough, K. (1996). Coupled Phase Theory for Sound Propagation in Emulsions. *Private Communication.*

Evans, T. (1986). Wave Propagation in Tissues. In *Physics in Medical Ultrasound* (Editor: Evans, T.), Institute of Physical Sciences in Medicine, U.K., IPSM Report No. 47, pp. 12–19.

Ewing, M. B. (1993). Thermophysical Properties of Fluids from Acoustic Measurements. *Pure and Applied Chemistry* **65**, 907–912.

Fairley, P. (1992). *Ultrasonic Studies of Foods Containing Air.* PhD Thesis, Leeds University, U.K.

Fairley, P., McClements, D. J., and Povey, M. J. W. (1991). Ultrasonic Characterization of some Aerated Foodstuffs. *Proceedings of the Institute of Acoustics* **13**, 63–70.

Fedotkin, I., Sukauskas, V., Chaperon, M., Shnaider, V., and Klimenko, M. (1980). Propagation of Ultrasonic Waves in Sugar Solutions and Sugar Juices. *Pishchevaya Promyshlennost', Respublikanskii Mezhvedomstvennyi Nauchno tekhnicheskii Sbornik No* **26**, 33–37.

Fenner, D. (1984). Ultrasound Measurements of Polystyrene Solutions in the Binodal, Theta, and Good-Solvent Regions. *Journal of Chemical Physics* **81**, 5179–5188.

Fikioris, J. G., and Waterman, P. C. (1964). Multiple Scattering of Waves. II "Hole Corrections" in the Scalar Case. *Journal of Mathematical Physics* **5**, 1413–1420.

Fillery-Travis, A. J., Gunning, P. A., Hibberd, D. J., and Robins, M. M. (1992). Mechanisms of Flocculation in Oil-in-Water Emulsions. Differentiation by Creaming Behaviour. In *Gums and Stabilisers for the Food Industry 6* (Editors: Phillips, G. O., Williams, P. A., and Wedlock, D. J.), pp. 363–370.

Fillery-Travis, A. J., Gunning, P. A., Hibberd, D. J., and Robins, M. M. (1993). Coexistent Phases in Concentrated Polydisperse Emulsions Flocculated by Non-Adsorbing Polymer. *Journal of Colloid and Interface Science* **159**, 189–197.

Foldy, L. L. (1945). *Physics Review* **67**, 107.

Fox, R. (1996). Probing the Body with Doppler Ultrasound. *Hospital Physics* 85–90.

Freyer, E. B. (1931). Sonic Studies of the Physical Properties of Liquids. II. The Velocity of Sound in Solutions of Certain Alkali Halides and Their Compressibilities. *Journal of the American Chemical Society* **53**, 1313–1320.

Frindi, M., Michels, B., and Zana, R. (1991). Ultrasonic Absorption Studies of Surfactant Exchange Between Micelles and Bulk Phase in Aqueous Micellar Solutions of Nonionic Surfactants with Short Alkyl Chains. 1. 1,2-Hexanediol and 1,2,3-Octanetriol. *Journal of Physical Chemistry* **95**, 4832–4837.

Frindi, M., Michels, B., and Zana, R. (1994a). Ultrasonic-Absorption Studies of Surfactant Exchange Between Micelles and the Bulk Phase in Aqueous Micellar Solutions of Amphoteric Surfactants. *Journal of Physical Chemistry* **98**, 6607–6611.

Frindi, M., Michels, B., Levy, H., and Zana, R. (1994b). Alkanediyl-Alpha,Omega-Bis(Dimethylalkylammonium Bromide) Surfactants. 4. Ultrasonic-Absorption Studies of Amphiphile Exchange Between Micelles and Bulk Phase in Aqueous Micellar Solutions. *Langmuir* **10**, 1140–1145.

Galema, S. A., and Høiland, H. (1991). Stereochemical Aspects of Hydration of Carbohydrates in Aqueous-Solutions. 3. Density and Ultrasound Measurements. *Journal of Physical Chemistry* **95**, 5321–5326.

Garnsey, R., Boe, R. J., Mahoney, R., and Litovitz, T. A. (1969). Determination of Electrolyte Apparent Molal Compressibilities at Infinite Dilution Using a High-Precision Ultrasonic Velocimeter. *Journal of Chemical Physics* **50**, 5222–5228.

Gaunaurd, G. C., and Überall, H. (1981). Resonance Theory of Bubbly Liquids. *Journal of the Acoustical Society of America* **69**, 362–370.

Gaunaurd, G. C., and Wertman, W. (1990). Comparison of Effective Medium and Multiple Scattering Theories of Predicting the Ultrasonic Properties of Dispersions: A Reexamination of Results. *Journal of the Acoustical Society of America* **87**, 2246–2247.

Gavish, B., Gratton, E., and Hardy, C. J. (1983). Adiabatic Compressibility of Globular Proteins. *Proceedings of the National Academy of Sciences of the U.S.A.* **80**, 750–754.

Gekko, K., and Hasegawa, Y. (1989). Effect of Temperature on the Compressibility of Native Globular Proteins. *Journal of Physical Chemistry* **93**, 426–429.

Gekko, K., and Noguchi, H. (1979). Compressibility of Globular Proteins in Water at 25°C. *Journal of Physical Chemistry* **83**, 2706–2714.

Gekko, K., and Yamagami, K. (1991). Flexibility of Food Proteins as Revealed by Compressibility. *Journal of Agricultural and Food Chemistry* **39**, 57–62.

Gladwell, N. R., Javanaud, C., Peers, K. E., and Rahalkar, R. R. (1985). Ultrasonic Behaviour of Edible Oils: Correlation with Rheology. *Journal of the American Oil Chemists' Society* **62**, 1231–1236.

Glazov, V. M., and Kim, W. (1988). Analysis of Concentration-Dependence of the Ultrasound Propagation: Velocity in the Systems with Congruently Melting Compounds, Based on Laplace Relation Modification. *Doklady Akademii Nauk SSSR* **301**, 365–368.

Gorbunov, M. A., Koshkin, N. I., and Sheloput, D.V. (1966). Acoustical Properties of Molecular Crystals Near the Melting Point. *Soviet Physics—Acoustics (English Translation)* **12**, 20–24.

Gouw, T. H., and Vlugter, J. C. (1964). Physical Properties of Fatty Acid Methyl Esters: IV. Ultrasonic Sound Velocity. *Journal of the American Oil Chemists' Society* **41**, 524–526.

Gouw, T. H., and Vlugter, J. C. (1966). Ultrasound Sound Velocity in Lipids. *Analytica Chimica Acta* **34**, 175–184.

Gouw, T. H., and Vlugter, J. C. (1967). Physical Properties of Triglycerides III: Ultrasonic Sound Velocity. *Fette, Seifen, Anstrichmittel* **69**, 159–164.

Griffin, W. G., and Griffin, M. C. A. (1990). The Attenuation of Ultrasound in Aqueous Suspensions of Casein Micelles from Bovine-Milk. *Journal of the Acoustical Society of America* **87**, 2541–2550.

Gunasekaran, S., and Ay, C. (1994). Evaluating Milk Coagulation with Ultrasonics. *Food Technology* **48**, 74–78.

Gunning, P. A., Hibberd, D. J., Howe, A. M., and Robins, M. M. (1988). Gravitational Destabilization of Emulsions Flocculated by Non-Adsorbed Xanthan. *Food Hydrocolloids* **2**, 119–129.

Gunning, P. A., Hibberd, D. J., Howe, A. M., and Robins, M. M. (1989). Use of Velocity of Ultrasound to Monitor Gravitational Separation in Dispersions. *Journal of the Society of Dairy Technology* **42**, 70–79.

Gunstone, F. D., Harwood, J. L., and Padley, F. B. (1994). *The Lipid Handbook*. Chapman and Hall, London.

Habeger, C. C. (1982). The Attenuation of Ultrasound in Dilute Polymeric Fiber Suspensions. *Journal of the Acoustical Society of America* **72**, 870–878.

Hampton, L. D. (1967). Acoustic Properties of Sediments. *Journal of the Acoustical Society of America* **42**, 882–890.

Harker, A. H., and Temple, J. A. G. (1988). Velocity and Attenuation of Ultrasound in Suspensions of Particles in Fluids. *Journal of Physics D: Applied Physics* **21**, 1576–1588.

Harker, A. H., and Temple, J. A. G. (1992). Propagation and Attenuation of Ultrasound in Suspensions. In *Developments in Acoustics and Ultrasonics* (Editors: Povey, M. J. W., and McClements, D. J.), Institute of Physics, Bristol, U.K., pp. 19–32.

Harker, A. H., Schofield, P., Stimpson, B. P., Taylor, R. G., and Temple, J. A. G. (1991). Ultrasonic Propagation in Slurries. *Ultrasonics* **29**, 427–438.

Harned, H. S., and Owen, B. B. (1958). *The Physical Chemistry of Electrolytic Solutions*, Third Edition. Reinhold, New York.

Harries, C. J. (1985). *The Use of Ultrasound for Detecting Particles Suspended in Lubricant and Hydraulic Fluids*. PhD Thesis, University of London, U.K.

Hassun, S. (1988). Ultrasonic Study of Molecular Association of Aqueous-Solutions of Polyvinyl-Alcohol: A Method to Determine Molecular-Weight. *European Polymer Journal* **24**, 795–797.

Heasell, E. L., and Lamb, H. (1956). The Absorption of Ultrasonic Waves in a Number of Pure Liquids Over the Frequency Range 100 to 200 Mc/S, *Proceedings of the Physical Society*, **119**, 869–877.

Hessen, B (1931). The Social and Economic Roots of Newton's 'Principia'. In *Science at the Crossroads* (Editor: Bernal, J. D.), Frank Cass & Co. Ltd, U.K., 2nd Edition, pp. 147–212.

Hibberd, D. J., Howe, A. M., Mackie, A. R., Purdy, P. W., and Robins, M. M. (1986). Measurement of Creaming Profiles in Oil-in-Water Emulsions. In *Food Emulsions and Foams* (Editor: Dickinson, E.), Royal Society of Chemistry, Cambridge, U.K.

Hiki, Y., and Tamura, J. (1981). Ultrasonic-Attenuation in Ice Crystals Near the Melting Temperature. *Journal de Physique (Orsay, France)* **42**, 547–552.

Holmes, A. K., and Challis, R. E. (1993). Ultrasonic Scattering in Concentrated Colloidal Suspensions. *Colloids and Surfaces A: Physiochemical and Engineering Aspects* **77**, 65–74.

Holmes, A. K., Challis, R. E., and Wedlock, D. J. (1993). A Wide Bandwidth Study of Ultrasound Velocity and Attenuation in Suspensions: Comparison of Theory with Experimental Measurements. *Journal of Colloid and Interface Science* **156**, 261–268.

Holmes, A. K., Challis, R. E., and Wedlock, D. J. (1994). A Wide-Bandwidth Ultrasonic Study of Suspensions: The Variation of Velocity and Attenuation with Particle Size. *Journal of Colloid and Interface Science* **168**, 339–348.

Horak, G. (1972). Real-time Ultrasonic-Spectroscopy in Suspensions. *Acustica* **37**, 11–20.

Howe, A. M., and Robins, M. M. (1990). Determination of Gravitational Separation in Dispersions from Concentration Profiles. *Colloids and Surfaces: An International Journal* **43**, 83–94.

Howe, A. M., Mackie, A. R., Richmond, P., and Robins, M. M. (1985). Creaming of Oil-in-Water Emulsions Containing Polysaccharides. In *Gums and Stabilisers for the Food Industry*.

Howe, A. M., Mackie, A. R., and Robins, M. M. (1986). Technique to Measure Emulsion Creaming by Velocity of Ultrasound. *Journal of Dispersion Science and Technology* **7**, 231–243.

Hussin, A. B. B. (1982). *An Investigation Into the Use of Ultrasonics to Monitor Phase Changes and Dilatations in Fats and Oils*. PhD Thesis, University of Leeds, U.K.

Hussin, A. B. B., and Povey, M. J. W. (1984). A Study of Dilation and Acoustic Propagation in Solidifying Fats and Oils. II. Experiment. *Journal of the American Oil Chemists' Society* **61**, 560–564.

Hustad, G. O., Richardson, T., Winder, W. C., and Dean, M. P. (1970). Factors Affecting Sound Velocity in Fats and Oils. *Journal of Dairy Science* **53**, 1525–1531.

Iqbal, M., and Verrall, R. (1989). Apparent Molar Volume and Adiabatic Compressibility Studies of Aqueous Solutions of Some Drug Compounds at 25°C. *Canadian Journal of Chemistry* **67**, 727–735.

Ivnitskii, D., Priev, A., and Shilnikov, G. (1987). Ultrasonic Velocimetry of Immunoglobulins. *Soviet Physics—Acoustics (English Translation)* **33**, 390–392.

Javanaud, C., and Rahalkar, R. R. (1988). Velocity of Sound in Vegetable Oils. *Fat Science and Technology* **90**, 73–75.

Javanaud, C., and Robins, M. M. (1993). Ultrasonic Methods. In *Food Process Monitoring Systems* (Editors: Pinder, A. C., and Godfrey, G.), Blackie, London, pp. 129–153.

Javanaud, C., Lond, P., and Rahalker, R. R. (1986). Evidence for Sound Absorption in Emulsions Due to Differing Thermal Properties of the Two Phases. *Ultrasonics* **24**, 137–141.

Javanaud, C., Gladwell, N. R., Gouldby, S. J., Hibberd, D. J., Thomas, A., and Robins, M. M. (1991). Experimental and Theoretical Values of the Ultrasonic Properties of Dispersions: Effect of Particle State and Size Distribution. *Ultrasonics* **29**, 331–337.

Jeffrey, A. (1980). *Mathematics for Engineers and Scientists*. Nelson, London.

Jha, D. K., and Jha, B. L. (1986). Free Volumes of Some I and IIb Group Salt Solutions in Water from Ultrasonic Velocity. *Colloids and Surfaces: An International Journal* **14**, 44–49.

Jones, D. S. (1986). *Acoustic and Electromagnetic Waves*. Oxford University Press, U.K.

Juszkiewicz, A., and Antosiewicz, J. (1986). Ultrasonic Velocity Studies on Carbohydrates in Aqueous Ethanolic Solutions. *Zeitschrift für Physikalische Chemie (Leipzig), Neue Folge* **148**, 185–195.

Kaatze, U., Kuhnel, V., Menzel, K., and Schwerdtfeger, S. (1993). Ultrasonic Spectroscopy of Liquids. Extending the Frequency Range of the Variable Sample Length Pulse Technique. *Measurement Science and Technology* **4**, 1257–1265.

Kaulgud, M. V., and Dhondge, S. S. (1988). Apparent Molal Volumes and Apparent Molal Compressibilities of Some Carbohydrates in Dilute Aqueous Solutions at Different Temperatures. *Indian Journal of Chemistry, Section A: Inorganic Bio Inorganic Physical Theoretical and Analytical Chemistry* **27**, 6–11.

Kaye, G. W. C., and Laby, T. H. (1986). *Tables of Physical and Chemical Constants*, 15th Edition. Longman Science, U.K.

Keatings, S. E. and Povey, M. J. W. (1996). *Ultrasound Measurement of Temperature Induced Changes in Egg White Proteins*. BSc Thesis, University of Leeds, U.K.

Kelly, T., McClements, D. J., and Povey, M. J. W. (1990). Sensor for Solid Fat Determination and Use Thereof. GB Patent Application No. 9013530.2.

Kharakoz, D. P. (1991). Volumetric Properties of Proteins and Their Analogs in Diluted Water Solutions. 2. Partial Adiabatic Compressibilities of Amino-Acids at 15-70-Degrees-C. *Journal of Physical Chemistry* **95**, 5634–5642.

Kharakoz, D. P., and Sarvazyan, A. P. (1993). Hydrational and Intrinsic Compressibilities of Globular-Proteins. *Biopolymers* **33**, 11–26.

Khimunin, A. S. (1972). Numerical Calculation of the Diffraction Corrections for the Precise Measurement of Ultrasound Absorption. *Acustica* **27**, 173–181.

Khimunin, A. S. (1975). Numerical Calculations of the Diffraction Corrections for the Precise Measurement of Ultrasound Phase Velocity. *Acustica* **32**, 192–200.

Kitamura, H., Sigel, B., Machi, J., Feleppa, E. J., Sokilmelgar, J., Kalisz, A, and Justin, J. (1995). Roles of Hematocrit and Fibrinogen in Red-Cell Aggregation Determined by Ultrasonic Scattering Properties. *Ultrasound in Medicine and Biology* **21**, 827–832.

Kittel, C. (1971). *Introduction To Solid State Physics*. Wiley, New York, pp. 157–197.

Kleis, S., and Sanchez, L. (1991). Dependence of Sound-Velocity on Salinity and Temperature in Saline Solutions. *Solar Energy* **46**, 371–375.

Kloek, W. (1995). *Measuring Crystallization Kinetics of Emulsified Mixtures of Triglycerides Using Ultrasound*. Internal Report, University of Leeds, U.K.

Kumar, A. (1991). Thermodynamic Properties of Pentanol Deduced from Ultrasonic Velocity: Data of its Aqueous-Solutions. *Physics and Chemistry of Liquids* **23**, 87–92.

Kuo, H. L. (1971). Variation of Ultrasonic Velocity and Absorption with Temperature and Frequency in High Viscosity Vegetable Oils. *Japanese Journal of Applied Physics. Part 1. Regular Papers and Short Notes* **10**, 167–170.

Kuo, H. L., and Weng, J. S. (1975). Temperature and Frequency Dependence of Ultrasonic Velocity and Absorption in Sperm and Seal Oils. *Journal of the American Oil Chemists' Society* **52**, 167–168.

Kuster, G. T., and Toksöz, M. N. (1974a). Velocity and Attenuation of Seismic Waves in Two-Phase Media: Part 1. Theoretical Formulations. *Geophysics* **39**, 587–618.

Kuster, G. T., and Toksöz, M. N. (1974b).Velocity and Attenuation of Seismic Waves in Two-Phase Media: Part II. Experimental Results. *Geophysics* **39**, 607–618.

Kuttruff, H. (1991a). *Ultrasonics: Fundamentals and Applications*. Elsevier Science, Amsterdam, The Netherlands.

Kuttruff, H. (1991b). *Ultrasonics: Fundamentals and Applications*. Elsevier Science, Amsterdam, The Netherlands, pp. 269–296.

Kuttruff, H. (1991c). *Ultrasonics: Fundamentals and Applications*. Elsevier Science, Amsterdam, The Netherlands, pp. 297–324.

Kytömaa, H. K. (1995). Theory of Sound Propagation in Suspensions: A Guide to Particle Size and Concentration Characterization. *Powder Technology* **82**, 115–121.

Lamb, H. (1925). *The Dynamical Theory of Sound*. Reprinted 1960, Dover, New York.

Lamb, J. (1969). Viscoelastic Relaxation in Supercooled Liquids and Polymers. *Rheologica Acta* **8**, 428–429.

Laplace, P. S. (1816). On the Velocity of Sound Through Air and Water. *Annales de Chimie (Paris)* **3**, 238–241.

Larkin, J. A. (1975). Thermodynamic Properties of Aqueous Non-Electrolyte Mixtures 1. Excess Enthalpy for Water + Ethanol at 298.15 to 383.15 K. *Journal of Chemical Thermodynamics* **7**, 137–148.

Lax, M. (1951). *Reviews of Modern Physics* **23**, 287.

Lax, M. (1952). *Physics Review* **85**, 621.

Leighton, T. G. (1996). *The Acoustic Bubble*. Academic Press, San Diego.

Lindsay, R. B. (1973). *Acoustics: Historical and Philosophical Development*. Dowden, Hutchinson and Ross, Stroudsburg, PA.

Lloyd, P. (1967a). Wave Propagation Through an Assembly of Spheres. II. The Density of Single-Particle Eigenstates. *Proceedings of the Physical Society, London* **90**, 207–215.

Lloyd, P. (1967b). Wave Propagation Through an Assembly of Spheres. III. The Density of States in a Liquid. *Proceedings of the Physical Society, London* **90**, 217–231.

Lloyd, P., and Berry, M. V. (1967). Wave Propagation Through an Assembly of Spheres. IV. Relations Between Different Multiple Scattering Theories. *Proceedings of the Physical Society, London* **91**, 678–688.

Ma, Y., Varadan, V. K., and Varadan, V. V. (1990). Comments on Ultrasonic Propagation in Suspensions. *Journal of the Acoustical Society of America* **87**, 2779–2782.

Mahony, C. (1987). The Ultrasonic-Detection of Platelet Aggregates. *Thrombosis Research* **47**, 665–672.

Mahony, C., Elion, J. L., and Fischer, P. L. C. (1989). A Computerized Analysis of Platelet-Aggregation Detected by Ultrasound. *Thrombosis Research* **55**, 351–360.

Majumdar, S., Holay, S. H., and Singh, R. P. (1980). Adiabatic Compressibility and Solvation of Drag Reducing Polymers in Aqueous Solutions. *European Polymer Journal* **16**, 1201–1206.

Manetou, A. (1990). *Development of an Ultrasonic Sensor for Monitoring Air Content*, Report No. 667. Leatherhead Food RA, U.K.

Mansfield, P. B. (1971). A New Wide-line NMR Analyzer and its Use in Determining the Solid-Liquid Ratio in Fat Samples. *Journal of the American Oil Chemists' Society* **48**, 4–6.

Mason, W. P., Thurston, R. N. and Pierce, A. D. (1964–1992). *Physical Acoustics.* Academic Press, San Diego.

Matheson, A. J. (1971). *Molecular Acoustics.* Wiley, New York.

Maxwell, J. L., Kurtz, F. A., and Strelka, B. J. (1984). Specific Volume (Density) of Saccharide Solutions (Corn Syrups and Blends) and Partial Specific Volumes of Saccharide-Water Mixtures. *Journal of Agricultural and Food Chemistry* **32**, 974–979.

McClements, D. J. (1988). *The Use of Ultrasonics for Characterising Fats and Emulsions.* PhD Thesis, University of Leeds, U.K.

McClements, D. J. (1991). Ultrasonic Characterisation of Emulsions and Suspensions. *Advances in Colloid and Interface Science* **37**, 33–72.

McClements, D. J. (1992). Comparison of Multiple Scattering Theories with Experimental Measurements in Emulsions. *Journal of the Acoustical Society of America* **91**, 849–853.

McClements, D. J. (1994). Ultrasonic Determination of Depletion Flocculation in Oil-in-Water Emulsions Containing a Non-Ionic Surfactant. *Colloids and Surfaces A: Physiochemical and Engineering Aspects* **90**, 25–35.

McClements, D. J., and Fairley, P. (1991). Ultrasonic Pulse Echo Reflectometer. *Ultrasonics* **29**, 58–62.

McClements, D. J., and Fairley, P. (1992). Frequency Scanning Ultrasonic Pulse Echo Reflectometer. *Ultrasonics* **30**, 403.

McClements, D. J., and Povey, M. J. W. (1987). Solid Fat Content Determination Using Ultrasonic Velocity Measurements. *International Journal of Food Science and Technology* **22**, 419–428.

McClements, D. J., and Povey, M. J. W. (1988a). Comparison of Pulsed NMR and Ultrasonic Velocity Techniques for Determining Solid Fat Contents. *International Journal of Food Science and Technology* **23**, 159–170.

McClements, D. J., and Povey, M. J. W. (1988b). Ultrasonic Velocity Measurements in some Triglycerides and Vegetable Oils. *Journal of the American Oil Chemists' Society* **65**, 1787–1789.

McClements, D. J., and Povey, M. J. W. (1989). Scattering of Ultrasound by Emulsions. *Journal of Physics D: Applied Physics* **22**, 38–47.

McClements, D. J., and Povey, M. J. W. (1992). Ultrasonic Analysis of Edible Fats and Oils. *Ultrasonics* **30**, 383–388.

McClements, D. J., Dickinson, E., and Povey, M. J. W. (1990a). Crystallization in Hydrocarbon-in-Water Emulsions Containing a Mixture of Solid and Liquid Droplets. *Chemical Physics Letters* **172**, 449–452.

McClements, D. J., Fairley, P., and Povey, M. J. W. (1990b). Comparison of Effective Medium and Multiple Scattering Theories of Predicting the Ultrasonic Properties of Dispersions. *Journal of the Acoustical Society of America* **87**, 2244–2245.

McClements, D. J., Povey, M. J. W., Jury, M., and Betsanis, E. (1990c). Ultrasonic Characterization of a Food Emulsion. *Ultrasonics* **28**, 266–272.

McClements, D. J., Dickinson, E., Dungan, S. R., Kinsella, J. E., Ma, J., and Povey, M. J. W. (1993a). Effect of Emulsifier Type on the Crystallization Kinetics of Oil-In-Water Emulsions Containing a Mixture of Solid and Liquid Droplets. *Journal of Colloid and Interface Science* **160**, 293–297.

McClements, D. J., Povey, M. J. W., and Dickinson, E. (1993b). Absorption and Velocity Dispersion due to Crystallization and Melting of Emulsion Droplets. *Ultrasonics* **31**, 433–437.

Mclure, I., Barbarincastillo, J., Neville, J., and Pethrick, R. A. (1994). Ultrasonic Velocities, Specific Volumes, Isobaric Thermal Expansivities, Isothermal Compressibilities and Isochoric Thermal Pressure Coefficients for Liquid Tetramethylsilane from 224.86-K to 273.28-K. *Thermochimica Acta* **233**, 325–328.

Meeten, G. H., and Sherman, N. E. (1993). Ultrasonic Velocity and Attenuation of Glass Ballotini in Viscous and Viscoelastic Fluids. *Ultrasonics* **31**, 193–199.

Meister, R., and St. Laurent, R. (1960). Ultrasonic Absorption and Velocity in Water Containing Algae in Suspension. *Journal of the Acoustical Society of America* **32**, 556–559.

Miecznik, P. (1993). Investigations of Complexing in Aqueous-Solutions of Dimethyl-Sulfoxide and Zinc-Chloride by an Ultrasonic Spectroscopy Method. *Acustica* **78**, 36–45.

Mikhailov, I. G., and Shutilov, V. A. (1965). Nonlinear Acoustical Properties of Aqueous Electrolytic Solutions. *Soviet Physics—Acoustics (English Translation)* **10**, 385–389.

Miles, C. A., Shore, D., and Langley, K. R. (1990). Attenuation of Ultrasound in Milks and Creams. *Ultrasonics* **28**, 394–400.

Miles, C. A., Fursey, G. A. J., and Jones, R. C. D. (1985). Ultrasonic Estimation of Solid/Liquid Ratios in Fats, Oils and Adipose Tissue. *Journal of the Science of Food and Agriculture* **36**, 215–228.

Millero, F. J., Ward, G. K., and Chetirkin, P. V. (1977). Relative Sound Velocities of Sea Salts at 25°C. *Journal of the Acoustical Society of America* **61**, 1492–1498.

Millero, F. J., Lo Surdo, A., and Shin, C. (1978). The Apparent Molal Volumes and Adiabatic Compressibilities of Aqueous Amino Acids at 25°C. *Journal of Physical Chemistry* **82**, 784–792.

Minnaert, M. (1933). On Musical Air Bubbles and the Sounds of Running Water. *Philosophical Magazine* **16**, 235–248.

Mitaku, S., Kagayama, T., and Kataoka, R. (1985). Ultrasonic Properties of Proteins. *Japanese Journal of Applied Physics. Part 1. Regular Papers and Short Notes* **24**, 43–45.

Morse, P. M., and Ingard, K: U. (1968). *Theoretical Acoustics*. McGraw-Hill, London.

Nishi, R. Y. (1975). The Scattering and Absorption of Sound Waves by a Gas Bubble in a Viscous Liquid. *Acustica* **33**, 65–74.

Nölting, B. (1995). Relation between Adiabatic and Pseudoadiabatic Compressibility in Ultrasonic Velocimetry. *Journal of Theoretical Biology* **175**, 191–196.

Nomoto, O., and Kishimoto, T. (1957). Molecular Sound Velocity of Aqueous Sugar Solutions. *Journal of the Physical Society of Japan* **12**, 311.

Nomura, H., Kawaizumi, F., and Iida, T. (1987). Partial Specific Compressibility and its Relation to Solvation of Solute Based on Measurements of Suspension System. *Bulletin of the Chemical Society of Japan* **60**, 25–30.

O'Brien, R. W. (1988). Electro-Acoustic Effects in a Dilute Suspension of Spherical Particles. *Journal of Fluid Mechanics* **190**, 71–86.

O'Brien, R. W. (1990). The Electroacoustic Equations for a Colloidal Suspension. *Journal of Fluid Mechanics* **212**, 81–93.

O'Brien, R. W., Garside, P., and Hunter, R. J. (1994). The Electroacoustic Reciprocal Relation. *Langmuir* **10**, 931–935.

O'Brien, R. W., Cannon, D. W., and Rowlands, W. N. (1995). Electroacoustic Determination of Particle Size and Zeta Potential. *Journal of Colloid and Interface Science*.

Ogawa, T., Yasuda, M., and Mizutani, K. (1984a). Volume and Adiabatic Compressibility of Amino Acids in Urea-Water Mixtures. *Bulletin of the Chemical Society of Japan* **57**, 662–666.

Ogawa, T., Mizutani, K., and Yasuda, M. (1984b). The Volume, Adiabatic Compressibility, and Viscosity of Amino Acids in Aqueous Alkali-Chloride Solutions. *Bulletin of the Chemical Society of Japan* **57**, 2064–2068.

Ogushwitz, P. R. (1985a). Applicability of the Biot Theory. I. Low Porosity Materials. *Journal of the Acoustical Society of America* **77**, 429–451.

Ogushwitz, P. R. (1985b). Applicability of the Biot Theory. II. Suspensions. *Journal of the Acoustical Society of America* **77**, 441–452.

Onori, G. (1988). Ionic Hydration in Sodium Chloride Solutions. *Journal of Chemical Physics* **89**, 510–516.

Owen, B. B., and Simons, H. L. (1957). Standard Partial Molal Compressibilities by Ultrasonics. I. Sodium Chloride and Potassium Chloride at 25°C. *Journal of Physical Chemistry* **61**, 479–482.

Pal, R. (1994a). Comments on Hold-Up (Volume Fraction) Measurements in Liquid/Liquid Dispersions Using an Ultrasonic Technique. *Industrial and Engineering Chemistry Research* **33**, 744–747.

Pal, R. (1994b). Techniques for Measuring the Composition (Oil And Water Content) of Emulsions– A State of the Art Review. *Colloids and Surfaces A: Physiochemical and Engineering Aspects* **84**, 141–193.

Paladhi, R., and Singh, R. P. (1990). The Effect of Molecular Weight on Adiabatic Compressibility and Solvation Number of Polymer Solutions. *European Polymer Journal* **26**, 441–444.

Pandey, J. D., Mishra, R. L., and Bhatt, T. (1977). Interaction Studies in Ternary Liquid Mixtures Ultrasonically. *Acustica* **38**, 83–85.

Parker, A., Gunning, P. A., Ng, K., and Robins, M. M. (1995). How Does Xanthan Stabilise Salad Dressing? *Food Hydrocolloids* **9**, 333.

Pavlovskaya, G. E., McClements, D. J., and Povey, M. J. W. (1992a). A Preliminary Study of the Influence of Dextran on the Precipitation of Legumin from Aqueous Salt Solutions. *International Journal of Food Science and Technology* **27**, 629–635.

Pavlovskaya, G. E., McClements, D. J., and Povey, M. J. W. (1992b). Ultrasonic Investigation of Aqueous Solutions of a Globular Protein. *Food Hydrocolloids* **6**, 253–262.

Pavlovskaya, G. E., McClements, D. J., and Povey, M. J. W. (1992c). Ultrasonic Studies of Aqueous Salt Solutions of a Globular Protein. In *Developments in Acoustics and Ultrasonics* (Editors: Povey, M. J. W. and McClements, D. J.), Institute of Physics, Bristol, U.K., pp. 243–246.

Pellam, J. R., and Galt, J. K. (1946). Ultrasonic Propagation in Liquids. I. Application of Pulse Echo Technique to Velocity and Absorption Measurements at 15 Megacycles. *Journal of Chemistry and Physics* **14**, 608–614.

Perrin, B. (1981). Ultrasonic-Attenuation in Molecular-Crystals. *Physical Review B: Condensed Matter* **24**, 6104–6113.

Pethrick, R. A. (1983). Ultrasonic Studies of Macromolecules. *Progress in Polymer Science* **9**, 197–295.

Pierce, A. D. (1981a). *Acoustics: An Introduction to its Physical Principles and Applications.* McGraw-Hill, New York.

Pierce, A. D. (1981b) Temperature and Pressure Dependence of the Sound Velocity in Distilled Water. In *Acoustics: An Introduction to its Physical Principles and Applications.* McGraw-Hill, New York, p. 33.

Pierce, A. D. (1981c) *Acoustics: An Introduction to its Physical Principles and Applications.* McGraw-Hill, New York, pp. 34-36.

Pierce, A. D. (1981d) *Acoustics: An Introduction to its Physical Principles and Applications.* McGraw-Hill, New York p. 232.

Pilhofer, G. M., and McCarthy, M. J. (1993). Phase Separation in Optically Opaque Emulsions. *Journal of Food Engineering* **20**, 369–379.

Pillai, S., Natarajan, S., and Palanisamy, P. (1983). Free-Volume of Sodium-Chloride Solutions in Water Dimethyl Formamide: Mixtures from Ultrasonic Velocity. *Journal of the Acoustical Society of America* **73**, 1616–1618.

Pinfield, V. J. (1996) *Studies of Creaming, Flocculation and Crystallization in Emulsions: Computer Modelling and Analysis of Ultrasound Propagation.* PhD Thesis, Leeds University, U.K.

Pinfield, V. J., and Povey, M. J. W. (1997). Thermal Scattering Must Be Accounted for in the Determination of Adiabatic Compressibility. *Journal of Physical Chemistry* **101**, 1110–1112.

Pinfield, V. J., Dickinson, E., and Povey, M. J. W. (1994). Modelling of Concentration Profiles and Ultrasound Velocity Profiles in a Creaming Emulsion: Importance of Scattering Effects. *Journal of Colloid and Interface Science* **166**, 363–374.

Pinfield, V. J., Povey, M. J. W., and Dickinson, E. (1995). The Application of Modified Forms of the Urick Equation to the Interpretation of Ultrasound Velocity in Scattering Systems. *Ultrasonics* **33**, 243–251.

Pinfield, V. J., Povey, M. J. W., and Dickinson, E. (1996). Interpretation of Ultrasound Velocity Creaming Profiles. *Ultrasonics* **34**, 695–698.

Povey, M. J. W. (1984). A Study of Dilatation and Acoustic Propagation in Solidifying Fats and Oils. I. Theory. *Journal of the American Oil Chemists' Society* **61**, 558–559.

Povey, M. J. W. (1995). Ultrasound Studies of Shelf-Life and Crystallization. In *New Physico-Chemical Techniques for the Characterization of Complex Food Systems* (Editor: Dickinson, E.), Chapman and Hall, London, pp. 196–213.

Povey, M. J. W. (1996). Measuring Aggregation in Colloids Using Ultrasound Velocity and Attenuation. In *Food Colloids - Proteins, Lipids and Polysaccharides* (Editors: Dickinson, E., and Bergenståhal, B.), Royal Society of Chemistry, Cambridge, U.K.

Povey, M. J. W., and McClements, D. J. (1989). Ultrasonics in Food Engineering: Part I. Introduction and Experimental Methods. *Journal of Food Engineering* **8**, 217–245.

Pryor, A.W., and Roscoe, R. (1954). The Velocity and Absorption of Sound in Aqueous Sugar Solutions. *Proceedings of the Physical Society, London* **67**, 70–81.

Puskar, A. (1982). *The Use of High Intensity Ultrasonics*. Elsevier Science, Amsterdam, The Netherlands.

Rajendran, V. (1993). Internal-Pressure Studies of Binary-Mixtures of Higher Alcohols in Tea At 303.15-K. *Indian Journal of Pure and Applied Physics* **31**, 812–814.

Raman, S. (1982). Ultrasonic Velocity Studies of Phase Transition of Solute, While in Solution. *Acustica* **52**, 42–43.

Rao, C. R., Reddy, L. C. S., and Prabhu, C. A. R. (1980). Study of Adulteration in Oils and Fats by Ultrasonic Method. *Current Science* **49**, 185–186.

Rao, R. (1940). *Indian Journal of Physics* **14**, 619.

Riebel, U., and Löffler, F. (1989). On-line Measurement of Particle Size Distribution and Particle Concentration by Ultrasonic Spectrometry. *Chemical Engineering Technology* **19**, 433–438.

Roberts, K. J. (1996). Ultrasound Analysis of Particle Size Distribution. *Materials World* **January**, 12–14.

Robins, M. M., Javanaud, C., Gladwell, N. R., Gouldby, S. J., Hibberd, D. J., and Thomas, A. (1991). The Effects of Particle State and Polydispersity on the Ultrasonic Properties of Dispersions. *Proceedings of the Institute of Acoustics* **13**, 79–86.

Rokhlenko (Rozlenko), A. A. (1986). Wide-Range Meter for Measuring Average Particle Sizes in Emulsions and Suspensions. *Measurement Techniques (English Translation) (Izmeritel'nya Tekhnika)* **29**, 581–584.

Rossini, F. D., Pitzer, K. S., Arnet, R. L., Braun, R. M., and Pimentel, G. C. (1953). *Selected Values of Physical and Thermodynamic Properties of Hydrocarbons and Related Compounds*. CAB International and Association of Applied Biologists.

Ross-Murphy, S. B. (1995). Rheology of Biopolymer Solutions and Gels. In *New Physico-Chemical Techniques for the Characterization of Complex Food Systems* (Editor: Dickinson, E.), Chapman and Hall, London.

Rowe A. M., Jr., and Chou, J. C. S. (1970). Pressure-Volume-Temperature-Concentration Relation of Aqueous NaCl Solutions. *Journal of Chemical and Engineering Data* **15**, 61–66.

Sakurai, M., Nakajima, T., Komatsu, T., and Nakagawa, T. (1975). Apparent Molal Compressibility of Sodium Chloride in Water. *Chemistry Letters* **1**, 971–976.

Sakurai, M., Nakamura, K., and Takenaka, N. (1994). Apparent Molar Volumes and Apparent Molar Adiabatic Compressions of Water in Some Alcohols. *Bulletin of the Chemical Society of Japan* **67**, 352–359.

Sakurai, M., Nakamura, K., Nitta, K., and Takenaka, N. (1995). Sound Velocities and Apparent Molar Adiabatic Compressions of Alcohols in Dilute Aqueous-Solutions. *Journal of Chemical and Engineering Data* **40**, 301–310.

Sarvazyan, A. P. (1991). Ultrasonic Velocimetry of Biological Compounds. *Annual Review of Biophysics and Biophysical Chemistry* **20**, 321–342.

Sarvazyan, A. P., and Hemmes, P. (1979). Relaxational Contributions to Protein Compressibility from Ultrasonic Data. *Biopolymers* **18**, 3015–3024.

Sayers, C. M. (1984). Scattering of Ultrasound by Minority Phases in Polycrystalline Metals. *United Kingdom Atomic Energy Authority Harwell Reports*.

Schaafs, W. (1967). Sound Velocity in Binary Mixtures and Solutions. In *Landolt-Börnstein New Series* (Editors: Hellwege, K. H., and Hellwege, A.M.), Springer-Verlag, New York, Vol. 2, pp. 79–109.

Schröder, A., and Raphael, E. (1992). Attenuation of Ultrasound in Silicone-Oil-in-Water Emulsions. *Europhysics Letters* **17**, 565–570.

Scott, D. M., and Boxman, A. (1995). Ultrasonic Measurement of Submicron Particles. *Particle and Particle Systems Characterization* **12**, 269–273.

Seki, H., Granato, A., and Truell, R. (1956). Diffraction Effects in the Ultrasonic Field of a Piston Source and Their Importance in the Accurate Measurement of Attenuation. *Journal of the Acoustical Society of America* **28**, 230–238.

Self, G., Povey, M. J. W., and Wainwright, H. (1992). What do Ultrasound Measurements in Fruit and Vegetables Tell You? In *Developments in Acoustics and Ultrasonics* (Editors: Povey, M. J. W., and McClements, D. J.), Institute of Physics, Bristol, U.K.

Sheppard, T. J. (1994). Solid State Gas Metering - the Future. *Flow Measurement and Instrumentation* **5**, 103–106.

Shiio, H. (1958). Ultrasonic Interferometer Measurements of the Amount of Bound Water. Saccharides. *Journal of the American Chemical Society* **80**, 70–73.

Shung, K., Krisko, B. A., and Ballard, J. O. (1982). Acoustic Measurement of Erythrocyte Compressibility. *Journal of the Acoustical Society of America* **72**, 1364–1367.

Slutsky, L. J. (1981). Ultrasonic Chemical Relaxation Spectroscopy in *Methods of Experimental Physics: Ultrasonics* (Edited by P. D. Edmonds, L. Marton, and C. Marton). Academic Press, San Diego, California.

Smith, D. E., and Winder, W. C. (1983). Effects of Temperature, Concentration and Solute Structure on the Acoustic Properties of Monosaccharide Solutions. *Journal of Food Science* **48**, 1822–1825.

Strutt, J. W. (Lord Rayleigh) (1872). Investigation of the Disturbance Produced by a Spherical Obstacle on the Waves of Sound. *Proceedings of the London Mathematical Society* **4**, 253–283.

Strutt, J. W. (Lord Rayleigh) (1877). *Theory of Sound*. Macmillan, London.

Strutt, J. W. (Lord Rayleigh) (1896). *The Theory of Sound*, 2nd Edition. Macmillan, London.

Suga, N. (1990). Biosonar Computation in Bats. *Scientific American* **6**, 34–41.

Tamura, J., Kogure, Y., and Hiki, Y. (1986). Ultrasonic-Attenuation and Dislocation Damping in Crystals of Ice. *Journal of the Physical Society of Japan* **55**, 3445–3461.

Tamura, J., Suzuki, K., and Mihashi, K. (1993). Adiabatic Compressibility of Myosin Subfragment-1 and Heavy-Meromyosin With or Without Nucleotide. *Biophysical Journal* **65**, 1899–1905.

Taylor, J. G. (1970). *Quantum Mechanics: An Introduction*. Unwin, London.

Taylor, J. R. (1982). *An Introduction to Error Analysis*. University Science Book, Mill Valley, Colorado.

Tiddy, G. J. T., Walsh, M. F., and Wyn-Jones, E. (1982). Ultrasonic Relaxation Studies of Concentrated Surfactant Solutions and Liquid Crystals. *Journal of the Chemical Society, Faraday Transactions* **78**, 389–401.

Tolley, J. A., and Rassing, J. (1983). A New Ultrasonic Velocity Scanning Technique Giving Concentration: Distance Profiles and Diffusion-Coefficient of Drugs in Gels. *International Journal of Pharmaceutics* **14**, 223–230.

Trusler, J. P. M. (1991). *Physical Acoustics and Metrology of Fluids*. Adam Hilger, Bristol, U.K.

Tsang, L., Kong, J. A., and Hashaby, T. A. (1982). Multiple Scattering of Acoustic Waves by Random Distribution of Discrete Spherical Scatterers with Quasicrystalline and Percus Yevick Approximation. *Journal of the Acoustical Society of America* **71**, 2779.

Tschoegl, N. W. (1989). *The Phenomenological Theory of Linear Viscoelastic Behaviour, an Introduction*. Springer-Verlag, New York.

Tsouris, C., and Tavlarides, L. L. (1994). Hold-Up (Volume Fraction) Measurements in Liquid—Liquid Dispersions Using an Ultrasonic Technique—Response. *Industrial and Engineering Chemistry Research* **33**, 748–749.

Twersky, V. (1978). Acoustic Bulk Parameters in Distributions of Pair-Correlated Scatterers. *Journal of the Acoustical Society of America* **64**, 1710.

Urick, R. J. (1947). A Sound Velocity Method for Determining the Compressibility of Finely Divided Substances. *Journal of Applied Physics* **18**, 983–987.

Urick, R. J., and Ament, W. S. (1949). The Propagation of Sound in Composite Media. *Journal of the Acoustical Society of America* **21**, 115–119.

Varadan, V. K., Varadan, V. V., and Ma, Y. (1985). A Propagator Model for Scattering of Acoustic Waves by Bubbles in Water. *Journal of the Acoustical Society of America* **78**, 1879–1881.

Waddington, D. (1986). Applications of Wide-Line NMR in the Oils and Fats Industry. In *Analysis of Oils and Fats* (Editors: Hamilton, R. J., and Rossel, J. B.), pp. 341–400.

Wade, T., Beattie, J. K., Rowlands, W. N., and Augustin, M. (1995). Electroacoustic Determination of Size and Zeta Potential of Casein Micelles in Skim Milk. *Journal of Dairy Research*.

Walker, R. C., and Bosin, W. A. (1971). Comparison of SFI, DSC and NMR Methods for Determining Solid-Liquid Ratios in Fats. *Journal of the American Oil Chemists' Society* **48**, 50–53.

Wang, Z., and Nur, A. J. (1991). Ultrasonic Velocities in Pure Hydrocarbons and Mixtures. *Journal of the Acoustical Society of America* **89**, 2725–2730.

Waterman, P. C., and Truell, R. (1961). Multiple Scattering of Waves. *Journal of Mathematical Physics* **2**, 512–537.

Watson, J. D., and Meister, R. (1963). Ultrasonic Absorption in Water Containing Plankton in Suspension. *Journal of the Acoustical Society of America* **35**, 1584–1589.

Weast, R. C. (1988a). In *Handbook of Chemistry and Physics* (Editor: Weast, R. C.), CRC Press, Boca Raton, Florida, 68th Edition, Table D228.

Weast, R. C. (1988b). Physical Constants of Clear Fused Quartz. In *Handbook of Chemistry and Physics* (Editor: Weast, R. C.), CRC Press, Boca Raton, Florida, 68th Edition, p. F58.

Wedlock, D. J., Fabris, I. J., and Grimsey, J. (1990). Sedimentation in Polydisperse Particulate Suspensions. *Colloids and Surfaces: An International Journal* **43**, 67.

Wedlock, D. J., McConaghy, C. J., and Hawksworth, S. (1993). Automation of Ultrasound Velocity Scanning for Concentrated Dispersions. *Colloids and Surfaces A: Physiochemical and Engineering Aspects* **77**, 49–54.

Wilson, W. D. (1959). Speed of Sound in Distilled Water as a Function of Temperature and Pressure. *Journal of the Acoustical Society of America* **31**, 1067–1072.

Wokke, J. M. P., and van den Wal, T. (1991). Application of Ultrasonic Waves to Quality Control During the Manufacture of Fat Containing Products. *Fat Science and Technology* **93**, 137–141.

Wood, A. B. (1941a). *A Textbook of Sound*. Bell and Sons, London.

Wood, A. B. (1941b). *A Textbook of Sound*. Bell and Sons, London. pp. 360–361.

Wood, A. B. (1964). *A Textbook of Sound*, 3rd Edition. Bell and Sons, London.

Wright, T., and Campbell, D. D. (1977). A UHF Ultrasonic Interferometer. *Journal of Physics E: Scientific Instruments* **10**, 1241–1244.

Yan, Y. D., and Clarke, J. H. R. (1989). In-Situ Determination of Particle Size Distributions in Colloids. *Advances in Colloid and Interface Science* **29**, 277–318.

Yasunaga, T., Usui, I., Iwata, K., and Miura, M. (1964). Ultrasonic Studies of the Hydration of Various Compounds in an Ethanol-Water Mixed Solvent. II. The Hydration of Organic Compounds. *Journal of the Chemical Society of America* **37**, 1658–1660.

Zemansky, M. W. (1957a). *Heat and Thermodynamics*. McGraw-Hill, New York.

Zemansky, M. W. (1957b). *Heat and Thermodynamics*. McGraw-Hill, New York, pp. 130–134.

Zemansky, M. W. (1957c). *Heat and Thermodynamics*. McGraw-Hill, New York, p. 260.

Ziman, J. M. (1966). Wave Propagation Through an Assembly of Spheres. I. The Greenian Method of the Theory of Metals. *Proceedings of the Physical Society, London* **88**, 387–405.

INDEX

16 bit, 167
32 bit, 167

absorption, 9
accuracy, 14, 15, 18, 25, 30, 31, 39, 142, 147–149, 171, 172, 174
   of solid fat content measurement, 68
acid gelation, 159
acoustic
   amplitude, 16
   energy, 173
   intensity, 112
   power, 112
Acoustic Time of Flight Measurement, 11
acoustical impedance, 16, 20
Acoustosizer, 150, 151
adiabatic, 3, 7, 91, 98, 100, 101, 107, 109, 126, 128, 129
   bulk modulus, 26
   compressibility, 2, 7, 26, 54
adulteration, 51
aggregation, 90, 94, 103, 129, 141, 143, 153, 154, 156–162
air, 11, 19, 28, 29
alcohol, 50
d'Alembert, 2
algae, 54
algorithmic, 165
alkyl halides, 45
amino acids, 45

amphiphilic surfactant, 84
amplitude, 16, 18, 20, 22–24, 33
   of the sound wave, 112
   factor, 112
analysis of errors, 15
apparent
   molar adiabatic compressibility, 38
   molar quantities, 37
   quantities, 39
   specific volume, 37
   yield stress, 152
aqueous solutions, 49
Aristotle, 1
asphaltene, 143, 157
associated Legendre polynomials, 97
AToM, *see* acoustic time of flight measurement
attenuation, 22, 24, 33, 54, 56, 57, 71, 79, 83, 92, 95, 96, 98–101, 104, 111, 113, 116, 118–122, 124, 127, 130, 131, 133, 134, 138–140, 142, 143, 145, 147, 148, 153, 154, 157–159, 161, 162, 172
   excess, 116
automate
   ultrasound profiling, 78
automatic control, 174
Avogadros principle, 35

backing, 14
backscatter, 2
backscattering, 140

bacteria, 74
  suspensions, 74
bandwidth, 147, 154, 168, 170, 172
bat, 2
beer foam, 163
Bentham, 169
Bessel functions, 111
binary mixture, 34, 41, 42
biological substances, 44
Biot
  coupled phase theory, 154
  theory, 134
blast, 138
  wave, 138
blend, 65, 71
block copolymer, 142
blood, 156, 161
blood cells
  adiabatic compressibility, 54
Boethius, 1
Bond
  acoustical, 19
bonding, 19, 158
boundary conditions, 97, 105, 112–114
boundary layer thickness, 127
broad band
  output, 173
  ultrasonic, 24
Brownian motion, 160
bubble
  resonance, 138
  size, 163
bubbles, 6, 28, 29, 48, 100, 137–140, 142
  microbubbles, 29
  resonant, 137
bubbly, 34
  liquids, 6
buffer rods, 18
bulk modulus, 26, 107–109, 153
  time-dependent, 107
burst rf, 147, 173

cabling
  faults, 173
calibration, 15, 25, 31, 49, 50, 79, 147, 172, 174, 175
carousel, 79
carrot cells, 54
casein, 70, 84, 88, 154, 157–159, 161
  α-, 159
  β-, 159
  κ-, 159

cavitation, 8, 20
  threshold, 20
cell
  bonding, 54
  suspensions, 54
  turgor, 54
cells, 54, 79
centrifugation, 48, 65
change of phase, see phase transition
characteristic impedance, 16
charged particles, 150
cheese, 159
chemical
  quantities, 35
  relaxation, 144
chocolate, 65
Chrysippus, 1
classical mechanics, 5
cleaning bath, 29
close packing, 142
cloud
  of bubbles, 137, 140
CMC, see critical micelle concentration
cooperative multitasking, 167
coalescence, 70, 72
cocoa, 65, 67, 68
  butter, 65, 67, 68
code, 165, 167–169, 175
coefficient of volume expansivity, 109
coherence, 6, 22
collapse, see gel
collision rate
  between droplets, 74
colloid, 52, 59, 75, 99, 102
colloidal systems, 59
complex impedance, 19
composite construction, 15
composition, 49, 52, 61, 63, 76, 90
  and ultrasonic properties, 76
compressibility, 25–27, 31, 34, 36–42, 44, 54, 55, 58, 71, 76, 81, 92, 93, 100, 104, 107, 109, 113, 122, 124–126, 128–131, 133, 134, 141, 152, 159
  changes
    compared with density change in oils, 58
  mean, 26
compression, 81, 85
computer modeling, 85
computer models, 158
computer processing unit
  CPU, 166
concentrated systems, 152, 163

concentration, 2, 35, 36, 39, 40, 49–52, 67, 72, 76, 81, 84, 85, 88, 103, 116, 125, 127, 128, 130, 134, 140–142, 152, 157–162
 increments, 39
 of droplets, 52
condensation, 26
conformational dynamics of proteins, 44
connectors
 faults, 174
conservation of volume, 81
constitutive equation, 113
continuous wave, 143
 techniques, 135
conversion efficiencies, 19
corrosion, 2
cosmetic, 129
coupled phase theory
 Biot theory, 154
coupled wave, 95
CPU
 computer processing unit, 166
cream, 138, 141, 143, 158, 161, 162, 174
 whipped, 163
creaming, 77, 81–84, 87, 90, 102, 158, 163, 174
 computer modeling, 77
critical micelle concentration, 144
crystal, 52, 56, 57, 59, 61, 63, 68, 70, 72–75, 79, 94
crystalline, 52, 56
crystallization, 72–74, 99, 102
crystallizing
 solids, 57
 systems, 59
custard powder, 152

damped, 14, 23
 transducers, 171, 173
Degas, 48
decay length, 101, 110, 125–127
deformation history, 107, 113
denaturation
 thermal, 44
density, 3, 8, 25–27, 31–34, 36–39, 42, 91, 92, 100, 101, 104, 110, 112, 117, 123–125, 128–131, 137, 139, 152
 mean, 26
depletion flocculation, 162
destabilize, 88
destabilizing mechanisms
 in foams, 163
diatoms, 54

differential
 experimental techniques, 42
 measurements, 39
diffraction, 22, 100
 corrections, 148
diffusion, 77, 85, 102, 103, 159, 160
 wave propagation, 34
digital
 averaging, 147
 oscilloscope, 170
dilatation, 26
 method, 61
dilation, 26
dilatometer, 75
dilatometry, 70, 72
dilute limit, 39–42
dipolar, 113
dipole scattering, 97
disk storage, 165, 166
dispersion relation, 108, 153
 for bubbles, 140
dispersions, 34
disproportionation, 163
dissipation, 33
distearin, 62
Doppler, 152, 160
droplet, 52, 58, 60, 72, 74, 75, 81
drugs
 diffusion in gels, 77
duty cycle, 20
dynamic mobility, 150, 151

echo ranging, 2
edible fats, 52
egg protein, 44
Einstein notation, 105
elastic, 7, 8
 modulus, 153
elasticity, 26
electrical impedance, 16, 19
electrosonic
 amplitude, 143
 analysis, 150
electroacoustics, 150
electromagnetic, 4, 6, 92
electromechanical, 19
emulsion, 52, 54–56, 58, 60, 61, 66, 68–70, 72–77, 84–86, 88, 90, 96, 99, 103, 116, 118, 124, 125, 129, 131, 134, 143, 150, 157, 158, 161, 162, 171, 172
 inversion, 34, 75, 76
 stability, 76, 78

emulsions, 52, 54, 55, 60, 61, 69, 74–77, 84, 90
energy, 94, 95, 106, 109, 111–113, 116, 117, 137, 138
    conservation equation, 106
    internal, 106
ensemble
    averaging, 117
    of bubbles, 140
envelope
    pulse, 17, 24
enzymes, 159
equation of continuity, 106
equations of motion, 104–106, 108
equidensity dispersions, 124
    and scattering theory, 124
errors, 14, 15, 19, 22, 23, 25, 42
ESA, 143, 150
ethanol, 34, 36, 37, 40, 42, 50, 128
Euler, 1
evaporation, 49
event driven, 165, 169
Excel, 79, 165, 170, 174
excess attenuation, 116
exchange of surfactant monomer, 144
excitation, 14
exciting
    transducer, 149, 150, 173
excluded volume, 103, 110

fast Fourier transform, 147, 148, 169
fat crystallization, 56
fats
    common features, 65
fatty acid, 52
filtration, 65
floc, 84, 154, 158, 161–163
flocculation, 84, 103, 129, 154, 158, 162, 163
flow
    measurement, 2
    profile, 152
    rate, 152
fluid motion, 110
foam drainage, 163
foams, 163
fogs, 133
food emulsion, 129, 143
FORTRAN, 116, 165
forward scattering, 94, 95
Fourier
    transform, 104, 147, 161, 169, 170, 172
fourth power dependence
    Rayleigh scattering, 3

fractal, 159
    dimension, 159
fractionation, 72, 83, 162
Fraunhofer diffraction, 22
freezing, 55, 56, 58
frequency, 12, 18–20, 24, 26, 33
Fresnel diffraction, 22
fruit, 51, 54

gas, 148, 163
GDI
    graphical display unit, 165
gel, 77, 85, 86, 88, 153, 158–161
gelation, 153, 159–161
gels, 153, 160
general relativity, 5
geometric limit, 142
geometric scattering, 100
geophysics, 134
glass, 152
glucose, 51
GPIB, 165, 168
gradient operator, 106
graphical display interface
    GDI, 165
gravimetry, 163
group velocity, 24, 33

Hankel functions, 113
hard fats, 65
hardened palm oil, 65
hardness, 65
harmonic, 26
hearing, 5
heat, 159
    capacity, 5
    flow, 94
    flux, 114
heterogeneous nucleation, 73
hexadecane, 55, 56, 58, 96, 134, 157
hierarchy method, 117
Hildebrandt equation, 63
HPIB, see GPIB, IEEE
hydration, 34, 44
    of a protein, 44
    shell, 44
hydrophilic, 159
hydrophobic, 159
    interactions, 134

ice, 51, 54, 56, 59, 67, 72, 74, 75, 77
ice nucleator, 74, 75

ice cream, 163
ideal solubility equation, 64; see also Hildebrandt equation
IEEE, 164, 168, 169, 174
impedance, 16, 18–20, 147, 149, 151
  matching, 147
in-line measurement, 61, 70
incoherent, 144
inertia, 104, 106, 110, 123, 124, 131, 133, 150
  inertial limit, 124
infinite time, 103
inhomogeneous, 8
instantaneous
  density, 26
  volume, 26
interfaces, 109
interference, 171
interferometry, 148
interparticle forces, 160
interpenetrating systems, 134
inverse problem, 145
isomers
  triacylglycerol, 52
isothermal, 3, 7, 8, 91; see also adiabatic
  isothermal compressibility, 44

Lagrange, 1
Lamé constants, 108, 153
lanolin, 74, 75
Laplace
  equation, 26, 34
  pressure, 163
latex
  polystyrene, 134
lattice model, 158
lead zirconium titanate, 14
LeCroy, 147, 148, 169, 170
Legendre polynomials, 111
light, 5, 6
light scattered, 171
light scattering, 141–143, 145, 152, 162
limiting quantities, 39
linear
  approximation, 20, 26
  regression, 36, 37
liquid, 91, 94, 96, 99–101, 104, 106, 108, 109, 111, 118, 134, 137, 138, 140
  liquidlike behavior, 108
lithium niobate, 142
long wavelength limit, 94, 98, 99, 104, 113, 118, 119, 125, 126, 130, 131, 133
longitudinal, 3, 9

loss
  modulus, 159
low frequency behavior, 140
'low' volume fraction, 140

macro, 148, 165, 175
macromolecules, 34
magnetic flea, 67
manufacturers, 47
margarine, 61, 65, 118
mass fraction, 27, 36, 37
match
  acoustical, 16–18, 31
matching, 4, 147, 149
MathCad, 116, 130, 148, 165
mayonnaise, 94, 129
mean
  of velocity of sound, 79
measurement chain, 31
measurement of the velocity of sound, 11
medical
  applications, 2
melting, 56, 58, 60, 63, 65, 67–69, 71
  point, 60, 63, 65, 67, 70, 71
memory effects, 106
method of intercepts, 36, 37, 42
micelle, 75, 144, 154, 159, 161, 162
microemulsions, 76
microscopy, 163
Mie theory, 142
milk, 152, 156, 159
mineral oil, 54, 55, 65, 66, 70, 74, 84
Minnaert frequency, 138
mismatch, see impedance
missing data, 146
mixtures, 11, 26, 29, 33, 34, 42
mode, 92, 94, 96, 101, 104, 106, 109, 111, 113–118, 121, 123, 124, 137, 138
  acoustic, 106
  propagational, 106
  shear, 106
  thermal, 106
model-dependent
  particle sizing, 146
model-independent, 145
modified scattering coefficients, 129
  a_n, 120
modified Urick equation, 41, 42, 76, 77, 81, 100, 125–128, 133, 140, 162
molality, 35, 36, 38
molar volume, 34, 36, 37, 40, 42, 44
molarity, 35, 38

mole fraction, 37
molecular
 formula
  and ultrasound velocity, 52
 sound velocity, 34
 transitions, 44
molecule, 52, 75, 84
molecules, 9
 as scatterers, 128
molten globule
 protein, 44
monodisperse, 77, 83
monopolar, 113
mouse click, 165
multitasking, 165
multiphase, 90
multiple
 dispersed phase, 129
 linear regression, 36, 37
 regression, 58
 scattering, 81, 92, 96, 97, 103, 116, 117, 124, 127, 128, 130, 133, 134, 137, 138, 140–142, 153, 160, 161
Musical Air Bubbles, 138

N-value, 61
National Instruments, 165
Navier-Stokes, 98, 101, 106
near field, 20, 22
networks, 159, 175
neutron, 100
Newton, 1, 3, 5
 second law, 26, 106
Newtonian fluid, 108
NMR, 60, 61, 65, 68, 70–72, 141, 152
 comparison with ultrasound velocity method, 70
Nomoto relation, 34
non-Newtonian, 101, 152
nucleation, 60, 72–75
number density, 129, 140
 and volume fraction, 140
Numerical
 calculations, 118
 simulations, 159

oil, 52, 54–56, 58, 61–63, 65, 66, 68, 69, 71–74, 77, 84, 86, 88, 90, 129
 droplets, 158, 161, 162
oil-in-water, 54–56, 58, 61, 65, 66, 69, 70, 72, 73, 77, 84, 88, 134
oils, ' 52, 63, 65, 84, 90

optical, 92
 coatings, 18
 microscopy, 74
original
 density, 26
 volume, 26
oscilloscope, 147, 148, 169, 170, 172
osmotic potential, 54
Ostwald ripening, 163
ovalbumin
 thermal denaturation, 45
overlap, *see* scattering

palm oil, 61, 65, 66, 69, 70, 72–74
paraffin, 61–63, 65
parallelism, 18, 22
partial
 molar volume, 36, 37, 40, 42
 specific adiabatic compressibility, 40
 specific volume, 40
 volume, 36, 41
partial wave analysis, 96, 111
 PWA, 100
particle
 accelerations, 20
 displacement, 20, 26
 shape, 72
 size, 70, 72, 74–77, 80, 81, 83
  distribution, 72, 81, 83, 118, 127, 145, 152, 162
  model-dependent approach, 146
  model-independent approach, 145
  strategies for determining, 147
 sizing, 24, 92, 94, 104, 120, 135, 141
PC, 165, 168, 169
permittivity, 93
phase, 5, 6, 17, 18, 20, 22, 24, 27, 32, 33, 42, 92, 94–99, 101–103, 110, 111, 115, 116, 118, 119, 121–123, 125–130, 132
 change, 90, 102; *see also* phase transition of a wave, 94
 sensitivity, 6
 shifts, 17, 171, 172
 transition, 54–57, 70
 velocity, 24
 volume, 141, 163
phonons, 5
phospholipids, 67, 68
piezoelectric pushers, 149
piston, 104
 source, 4, 20
pitch and catch, 11

platinum resistance thermometer, 14
pNMR, 61
  direct method, 61
  weight method, 61
polydispersity, 83, 129, 162
polymer, 129, 142, 144, 161
  solutions, 144
polymers, 108
polymorphic form, 70, 72
polymorphism, 72
polynomial, 34, 37, 41
polysaccharide, 84
polystyrene, 96
  latex, 134
  spheres, 134
polyvinylidene difluoride, 142
POP, 61
POS, 61
potential
  scalar potential, 105
  vector, 112
  vector potential, 105
  wave potential, 112
power, 16, 19, 20, 29
  levels, 2, 8, 19, 20
power transfer, 16
preemptive multitasking, 167
precision, 14, 25, 31
  of solid fat content measurement, 70
pressure, 12, 16, 19, 20, 22, 26, 29, 31, 33, 95, 96, 99, 101, 105, 106, 108–110, 114, 122, 123, 128, 138
  hydrostatic, 106
Principia, 1
profiler, 161, 162, 165, 171, 173
profiles
  interpretation of ultrasound velocity, 80
profiling, 76, 77, 89, 90
  ultrasound, 157, 161, 162, 164, 165, 173, 174
propagation equations, 153
propagational mode, 106, 109, 111, 113, 116
protein, 159
  conformation, 57
proteins, 44, 75
protocol, 168
prototyping, 167
pulsation, 113, 138
pulse, 12, 13, 15, 17, 19, 20, 22–25, 31, 92, 102–104, 109, 110, 134, 135, 137
  amplitude, 169
  echo, 2, 8, 11

  overlap, 171
  reflectometer, 148, 163
  envelope, 17, 24
  methods, 149, 154
  shape, 169
  techniques, 24, 135
pulsed Nuclear Magnetic Resonance
  pNMR, 61, 70
PVDF, 142, 147

quantization of sound, 5
quantum
  mechanics, 5
  theory, 5
quarter wave, 93
  transformer, 18
quartz, 18, 142, 147
  velocity of sound in, 18

radar, 2
radial
  distribution function, 140
  velocity, 114
RAM, 165
  random access memory, 165
random errors, 14
ratio of specific heats, 2, 108
Rayleigh, 93, 96, 98
  scattering, 93, 162
  -Gans-Debye, 142
reactive
  component of impedance, 16
rectified diffusion, 20
reflection, 14, 16, 17
  coefficient, 16, 140, 163
  techniques, 163
reflectometer, 148, 163
refractive index, 51
relaxation, 8, 56, 57, 71, 144
  accounting for, 144
  dependence of attenuation on frequency, 145
  effects, 108
  frequency, see relaxation time
  time, 56, 71
rennet, 159
renormalization, 80–83, 162
repetition rate, 20
resistive
  component of impedance, 16
  temperature device, 14
resonance, 28, 138–140

resonant, 137
  frequency, 138
reverberation, 17, 18, 171
rheological, 88
rheology, 52
rigidity
  modulus, 153, 159
RS232, 164, 168
RTD, 14

SA, 62
salt, 65
sample
  changer, 79
  size, 173
scalar potential, 113
scaling factor, 118
scan
  ultrasound profiling, 81, 82
scattering, 3, 5, 6, 8, 10, 41, 58, 72, 76, 77, 80, 81, 125
  coefficient, 96, 102, 113, 115, 117, 118, 120–125, 127–131, 137, 140, 141, 145, 153
    single particle, 115, 118
  concentration dependent multiple scattering, 128
  experimental data, 118
  modified scattering coefficient, 127
  single particle
    for bubbles, 140
    by a bubble, 138
  strong scattering, 137
  theory, 58, 76, 81
  thermal, 125
  thermoelastic, 104
  viscoinertial, 104, 110
sedimentation, 77
seismic, 134
self-consistency, 99, 103, 104
sensitivity, 49, 71, 72
  of ultrasound
    method, 71
    velocity, 49
serial communications, 164, 168
serum, 77, 85, 86, 88, 90, 158, 159, 161, 171, 172
server, 175
SFC, 52, 61, 65, 69; see also solid fat content
shape, 72
  nonsphericity, 102
  particle, 127

shear, 92, 94, 99, 101, 103, 104, 106, 107, 109–116, 118, 123, 124, 126, 127
  mode, 4, 7, 9, 102
  modulus, 113; see also rigidity modulus
  thickening, 152
  wave, 160
shelf life, 90
signal-to-noise, 147
silver iodide, 74
size, 52, 70, 72, 74–77, 80, 81, 83, 94, 141–143, 145, 148, 150–152, 157–159, 161–164, 167, 169
  distribution, 145, 151, 152, 162, 163, 169
  of droplets, 52
  particle size, 128
    distribution, 118, 127
skimmed milk, 156, 159
skin depth, 110, 127
sky
  blue sky, 2, 3, 6, 7
SNOMAX, 74
sodium caseinate, 70, 84, 88
soft
  soft-solid, 11
solid, 94, 100, 102, 104, 106–109, 113, 118, 134, 137
  content, 56–59, 61, 63, 68–72
    very high, 71
    very low, 71
  fat content, 52, 55, 60, 61, 65, 67, 68, 70, 71, 90
  measurement of velocity of sound in, 65
  solidlike behavior, 107
solid–liquid transition, 71
solubility, 61, 63, 64
solute, 31, 34–40, 42, 44
solution–emulsion transition, 75
solutions, 34
  scattering in, 129
sonar, 138
sonication, 29
sound
  speed, see velocity of sound
  velocity, 40; see also velocity of sound
Sounds of Running Water, 138
specific heat, 34, 44, 76, 101, 109, 128
  ratio of specific heats, 108
spectrometry, 161
spectrum, 146
speed of sound, 30, 31
spherical Bessel functions, 97
spherical coordinate system, 111

spike
  method, 147, 171, 173
sponge, 134
spread sheet program
  the use of, 65
spreadsheet, 148, 165, 170, 174
SS, 62
stabilizer, 159
stability, 76, 78, 90
standard deviation, 79
stearic acid, 62
stepper
  motor, 79, 169, 174
stiffness, 26
stirring, 67, 75
  gentle, 75
  magnetic flea, 67
Stokes-Navier equation, 106
storage
  modulus, 153, 159, 165, 166, 174, 175
strain, 26, 106–108
strain-rate tensor, 106–108, 152, 153
stress, 26, 98, 102, 106–108, 113–115,
  152, 153, 161, 174
  principle stress, 108
  tensor, 106–108, 113
strong scatterer, 137
structure factor, 100
submarine, 138
submicelles, 154
sugar, 51, 67; *see also* carbohydrates
  concentration, 51
sunflower, 65, 84, 109, 118, 120, 131, 133
supercooling, 69, 73, 74
surface energy, 137
surface tension, 138
surfactants, 29, 48, 69, 84, 144
  characterization of, 144
  exchange, 144
syneresis, 89
systematic errors, 14

tangential velocity, 114
tasks, *see* multitasking
temperature, 11, 14, 15, 20, 25, 29, 31, 35,
  37, 44
  coefficient of the velocity of sound, 29
  coefficient of velocity, 11, 31
    negative, 11
  control, 79, 149, 174
  dependence, 71, 77, 143, 173
  differences, 49

  logging, 174
  measurement, 174
  for a profiling experiment, 77
  programmer, 74
  wave, 91
tempering, 65
text box, 165
thermal, 91, 92, 96, 99, 101–104, 106,
  109–111, 113–116, 118, 119, 122, 123,
  125–129, 131, 133; *see also* thermal decay
  length
  diffusivity
    thermometric conductivity, 109
  effects, 72
  expansivity, 76, 109, 114
  homogeneity, 75
  mode, 114
  properties, 76
  scattering, 42, 96
  stresses, 98, 102
  wave, 92, 101, 109, 110, 123, 128, 160
  wavelength, 123
thermoelastic, 131
  scattering, 104
thermodynamics, 2, 9, 44
  parameters, 34
thermometric conductivity, 109, 114
thin slab approximation, 116
time-of-flight, 13, 14, 22, 31, 169, 171, 172
timer, 13, 15
timer-counter, 169, 172
tissue
  medical applications, 2
tomography, 143
tone burst, 79, 80
Tr, 106
trace, 106
transducer, 14–20, 22, 23, 33, 70, 79, 142,
  147, 150–152, 170–174
  characterization, 172
  excitation, 172
transmission coefficient, 16
triacylglycerols, 52, 61, 62, 64; *see also*
  triglycerides
  isomers, 52
triangular wave, 18
trigger, 23, 79, 80, 165, 169, 171, 172
  errors, 80
  settings, 172
triglyceride
  triacylglycerol, 52, 61, 75
tuning, 147

turgor, 54
Tween, 154, 162
  diameter of Tween micelle, 162
Tween20, 69

Ultrasonic Sugar Index, 51
ultrasound
  bath, 8
  profiler
    the Leeds profiler, 78
  velocity
    meter, UVM, 173
    profiles, 76, 80
  technique
    advantages and disadvantages, 71
unicellular plant, 54
unreliability, 15
unsaturated bonds, 52
Urick equation, 27, 31, 34, 40–42, 49, 52, 55–58, 63, 76, 77, 81, 92, 124, 133, 162; see also modified Urick equation
UVM, 50, 74, 173

vector
  calculus, 104
  potential, 112
  quantity, 106
vegetables, 54
velocimeter, 13, 23
velocity
  acoustic, 105
  local velocity of the fluid continuum, 105
  low velocity in bubble suspensions, 140
  sound, 105
  ultrasound, 105
  variation with temperature, 52
  wave velocity, 93
velocity of sound, 1, 3, 8, 11, 12, 15, 19, 25–31, 34, 39
  in a cell, 54
  in vegetable oils, 52
  temperature coefficient, 54
  temperature dependence, 49
Vetruvius, 1
viscoelastic, 92, 94, 106, 108, 113, 152
viscoinertial, 104, 110, 123, 131, 133, 150
  scattering, 104, 123
viscoplastic, 152, 153
viscosity, 91, 92, 99, 108–110, 113, 150, 151, 153
viscous, 92, 101, 106, 109, 114, 123–127, 133
  decay length, 162
  samples, 48
Visual Basic, 79, 165
visual display unit, 165
volume expansivity, 11, 44, 109
  negative, 11
volume fraction, 41, 49, 50, 58, 61, 67, 74, 77, 81–84, 87, 88, 94
  and number density, 140

water, 11, 15, 17, 19, 20, 24, 25, 28–31, 33, 34, 36–38, 40, 42, 44, 49, 50, 54–56, 58, 61, 65, 66, 69–75, 77, 79, 84, 88, 90, 91, 96, 99–101, 103, 104, 109, 118, 120, 123, 126–129, 131, 133, 134, 137, 138, 143, 147, 157, 159, 162, 171–174
  content
    determination by velocity of sound, 54
  droplets, 99
  equation, 93
  number, 24
  phenomenon, 1
  vector, 33, 95, 99, 101
wavenumbers, 105
wave–particle duality, 5
waveform analysis, 169
waveforms, 14, 23
wavefront, 20, 22, 34
wavelength, 3–7, 9, 12, 18, 24, 33, 104
  long wavelength limit, 104
wavenumber, 83, 100, 101, 108, 116, 117
  effective, 116
wax, 129
weak scattering, 92, 97, 100, 104, 116, 137
wear plate, 14, 15, 23
whey solids, 65
Windows, 165–167, 175
Wood equation, 26, 31, 34, 81, 91, 101, 153

X-ray, 100
xanthan, 77, 84, 86, 87, 90

yield stress, 152, 161
yoghurt, 153, 163

zeta potential, 150, 151

CPSIA information can be obtained at www.ICGtesting.com
Printed in the USA
BVOW04*1301070913
330099BV00008B/262/P